Environmental Assessment in Practice

Environmental assessment (EA) has become established worldwide as an environmental management tool used by government agencies, companies and other organisations to identify, predict and evaluate the potential biological, physical, social and health effects of projects and other development actions. While improvements have been achieved, there remain many opportunities for strengthening the capacity for conducting EA studies and improving the design and implementation of EA systems.

Environmental Assessment in Practice introduces what constitutes good practice in EA and demonstrates the use of methods and techniques for impact identification, prediction and evaluation; environmental risk assessment; consultation and participation; project management; environmental statement review and post-project analysis; and strategic environmental assessment. A wide range of British and international case studies illustrates and explains how the different methods, techniques and disciplines of EA can be used in practice.

Growing environmental awareness and increasing public concerns over the impacts of developments on biophysical systems have made the use of environmental assessment systems invaluable. This book gives a comprehensive coverage of the subject and will be invaluable to practitioners and students alike.

D. Owen Harrop and **J. Ashley Nixon** have wide international experience in EA studies, lecturing and training; both are consultants for CORDAH Environmental Management Consultants, Aberdeen.

Routledge Environmental Management Series

This important series presents a comprehensive introduction to the principles and practices of environmental management across a wide range of fields. Introducing the theories and practices fundamental to modern environmental management, the series features a number of focused volumes to examine applications in specific environments and topics, all offering a wealth of real-life examples and practical guidance.

MANAGING ENVIRONMENTAL POLLUTION
Andrew Farmer

COASTAL AND ESTUARINE MANAGEMENT
Peter W. French

ENVIRONMENTAL ASSESSMENT IN PRACTICE
D. Owen Harrop and J. Ashley Nixon

Forthcoming titles:

WETLAND MANAGEMENT
L. Heathwaite

COUNTRYSIDE MANAGEMENT
R. Clarke

Environmental Assessment
in Practice

D. Owen Harrop and
J. Ashley Nixon

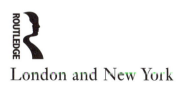

London and New York

First published 1999 by Routledge
11 New Fetter Lane, London EC4P 4EE

Simultaneously published in the USA and Canada
by Routledge
29 West 35th Street, New York, NY 10001

Typeset in Ehrhardt by Keystroke, Jacaranda Lodge, Wolverhampton
Printed and bound in Great Britain by TJ International, Padstow,
Cornwall

British Library Cataloguing in Publication Data
A catalogue record for this book is available from the British Library

Library of Congress Cataloging in Publication Data
Harrop, D. Owen, 1956–
    Environmental assessment in practice / D. Owen Harrop and J. Ashley Nixon.
      p.   cm. — (Routledge environmental management series)
    Includes bibliographical references and index.
    1. Environmental risk assessment. 2. Environmental management
I. Nixon, J. Ashley, 1957–. II. Title. III. Series.
GE145.H37   1999
363.7′02—dc21   98–20257

ISBN 0–415–15690–4 (hbk)
ISBN 0–415–15691–2 (pbk)

Dedication

This book is dedicated to Callum, Jenny, Matthew, Sarah and William in the hope that they will become tomorrow's environmentalists.

The authors would like to make a special dedication to Robert Turnbull (1919–1998), a close and admired colleague.

Contents

List of figures   x
List of tables   xi
List of case studies   xiii
Preface   xv
Acknowledgements   xvi

1 Introduction to environmental assessment:
   purpose and procedures   1

   Introduction   2
   Terminology   3
   General principles of EA   3
   International developments in EA   5
   Costs and benefits of EA   7
   Overview of the main stages in the EA process   9
   Questions for thought   13

2 Environmental assessment methods   15

   Introduction   16
   Baseline studies   18
   EA identification methods   18
   Questions for thought   32

3　Techniques for impact prediction and evaluation　33

Introduction　34
Air quality assessment　34
Noise assessment　39
Landscape and visual impact assessment　45
Ecological assessment　52
Water assessment　58
Archaeological and cultural heritage assessment　64
Social impact assessment　66
Questions for thought　71

4　Environmental risk assessment　73

Introduction　74
Terminology　75
Applications of risk assessment　79
Questions for thought　88

5　Consultation and participation: the public role in　89
environmental assessment

Introduction　90
Public participation in EA in Europe and the UK　90
Formal and informal opportunities for public participation
　in the EA process in the UK　91
Strengths and weaknesses of formal and informal public participation
　in the UK　95
Real and perceived barriers to public participation in the UK　97
Future trends and mechanisms for the promotion of public
　participation in EA　99
Questions for thought　107

6　Managing the EA process　109

Introduction　110
Context and procedure　111
Technical management　111
Report writing　116
Financial control　119
Questions for thought　125

7   Quality assurance in EA: ES review and   127
     post-project analysis

     Introduction   128
     Reviewing ESs   128
     ES review in the UK   129
     International ES review procedures   134
     Post-project analysis: auditing and monitoring in EA   138
     Conducting a post-project analysis   139
     Questions for thought   146

8   Strategic environmental assessment   149

     Introduction   150
     Benefits of SEA   150
     Assessment of cumulative effects   151
     Comparison of SEA and project-level EA   152
     Key tasks and activities   155
     Using SEA   157
     Questions for thought   169

9   EA in practice   171

     Introduction   172
     Example 1: Chipboard manufacturing plant   172
     Example 2: Pig breeding centre   184

     The final word   200

     Further reading   201

     References   202
     Index   216

Figures

1.1   A generalised procedure for EA   8
1.2   Two-stage screening procedure   11
2.1   Direct and indirect impacts   17
2.2   Matrix of a proposed power plant   22–3
2.3   Matrix for the determination of alternative project sites   24
2.4   Simplified network for air-quality issues for a proposed incinerator   26
2.5   Application of environmental features mapping   27
3.1   AQA procedures   36
3.2   Dispersion coefficients σ_y and σ_z 39
3.3   Site location map   50–1
3.4   First flush effect of stormwater discharges   63
3.5   Archaeological assessment procedure to evaluate the severity of impact   65
4.1   QRA procedural steps   75
5.1   UK EA procedures   92
6.1   QA EIA costings format for project expenditure   121
6.2   Assessment methodology for sewage sludge disposal options   126
7.1   Hierarchical structure of the Manchester ES review package   130
7.2   Predicted and actual impacts associated with the construction
and operation of the Greater Manchester Metrolink (Phase 1)   143
8.1   Potential pathways leading to cumulative environmental impacts   152
8.2   Sequence of actions and assessments (SEA and EA) within
a comprehensive planning and assessment system   153
8.3   SEA for the Lake Myvatn area, Iceland   164
9.1   Site location map   173
9.2   Site location map   186

Tables

2.1   Example of information included in a descriptive checklist   20
2.2   Scoping matrix used for the proposed third River Don
       crossing, Aberdeen   29
2.3a Descriptive matrix used for land drainage schemes:
       construction phase   30–1
2.3b Descriptive matrix used for land drainage schemes: end state phase   31
3.1   Pasquill stability categories   40
3.2   Fitted constants for the Pasquill diffusion parameters   40
3.3   Field survey sheet for landscape assessment   47
3.4   Landscape and visual sensitive receivers   49
3.5   Plant communities of the National Vegetation Classification   54
3.6   River corridor survey zones   55
3.7   Ecological assessment and reinstatement recommendations
       for a site forming part of the North Western Ethylene Pipeline   58
4.1   Risk assessment criteria   77
4.2   QRA predictions for proposed incineration plants in the UK   78
4.3   Qualitative assessment matrix for groundwater receptors   80
4.4   Maximum long-term annual ground-level concentrations and
       recommended air-quality guidelines for a proposed waste to
       energy plant   81
4.5   QRA data assumptions   83
4.6   Predicted lifetime risk of developing cancer   84
4.7   Estimated soil concentrations resulting from deposition of maximum
       predicted long-term ground-level air concentrations   85
6.1   Review questions to be asked prior to undertaking the EA
       management process   112

6.2	Specialists likely to be used in EAs	113
6.3	Generic topical outline for an ES	118
6.4	Cost of undertaking EAs	120
6.5	Layout of ES for proposed sewage sludge incinerator	124–5
7.1	ES review criteria	131–3
7.2	Number of impact predictions for each case study	141
8.1	Some basic steps in conducting an SEA	156
8.2	Scenarios for annual visitor numbers to the Victoria Falls area	169
9.1	Summary of peak hour vehicle movements	180
9.2	Comparison of predicted ground-level concentrations from the proposed plant with recommended air-quality criteria	183
9.3	Inputs and outputs of livestock and materials	189

Case studies

River Don Crossing, Aberdeen   28
Alder Road Weir   30
Air dispersion modelling scoping exercise   38
Noise assessment of an industrial plant   44
Assessment of a proposed chipboard manufacturing plant   49
North Western Ethylene Pipeline ecological assessment   57
Urban stormwater runoff assessment   63
Archaeological assessment procedure   66
Guri hydroelectric power project, Venezuela   69
QRA of a waste to energy plant   80
Gold mining in Connemara and South Mayo, Ireland   101
Sand and gravel extraction, Bedfordshire, UK   102
The Fylde Forum   103
Testwood Lakes   104
South Warwickshire Prospect   105
Waste management in Alberta, Canada   105
Project management of an EA for a sewage sludge incinerator   123
Audit study of four developments in the UK   141
The Greater Manchester Metrolink scheme   142
Impacts-backwards audit of a small surface coal mine   144
Environmental appraisal of the Kent Structure Plan   161
Environmental appraisal of the Grampian Structure Plan   162
SEA for the Lake Myvatn area in Skutustadahreppur, Iceland   163
SEA for the Victoria Falls area, Zambia/Zimbabwe   166

Preface

With over twenty-five years of experience and over 200 systems in place, environmental assessment is internationally well established as a part of development planning, used by government agencies, companies and other organisations. During this time, many thousands of environmental studies have been prepared and a considerable literature on the subject has been written. As more people become engaged in the various aspects of the environmental assessment process, be they planning officers, environmental consultants or members of private or non-governmental organisations, there continues to be a demand for information and instruction on how to conduct an environmental assessment study. This practical perspective forms the basis of this book and the authors have sought to explain what constitutes good practice in applying environmental assessment as an environmental management tool. The book does not attempt to provide a theoretical grounding in the subject, nor any detailed analysis of environmental assessment policy and legislation as this has been well covered in a number of recent publications. Instead, particular emphasis is given to case studies, some of which were contributed to by the authors themselves, to show how the different methods, techniques and disciplines of environmental assessment can be used. While the book presents examples and case studies from a predominantly British context, it is hoped that it is of value to readers in other countries who wish to work within this important part of environmental management.

Acknowledgements

The authors are grateful to many people who helped in various ways in the preparation of this book and those who gave permission to cite their work. Some of the materials used were originally developed for various international training courses in environmental assessment and management organised by the Centre for Environmental Management and Planning (CEMP). For this we extend our appreciation to our colleagues who helped to organise and present these courses and in particular the contributions of Charlotte Bingham, Clare Brooke, Larry Canter, Brian D. Clark, Stefan Gorzula, Ron Bisset, Jennifer McQuaid-Cook, Barry Sadler, Deborah Stephenson, Fiona Walsh, Simon Pollard and Lee Wilson.

Robert Turnbull, Larry Canter, Jeff Obbard and John Boldon are especially thanked for their helpful comments at the drafting stage of the text. Doug Reid is thanked for his assistance in the preparation of the figures.

Finally we are grateful to our colleagues at Cordah Environmental Management Consultants, Aberdeen, and staff at Aspinwall & Company for their support and encouragement and for contributing information.

While environmental statements are public documents, the authors have obtained, where possible, permission to use materials from the consultancy which produced the document. The opinions relating to the case studies cited are, however, those of the authors and do not necessarily reflect the findings of the original studies.

Every effort has been made to contact copyright holders and we apologise for any inadvertent omission. If any acknowledgement is missing it would be appreciated if contact could be made care of the publishers so that this can be rectified in any future edition.

Introduction to environmental assessment

Purpose and procedures

◆ Introduction   2
◆ Terminology   3
◆ General principles of EA   3
◆ International developments in EA   5
◆ Costs and benefits of EA   7
◆ Overview of the main stages in the
  EA process   9
◆ Questions for thought   13

Introduction

Since the enactment of the National Environmental Policy Act (NEPA) in the USA in 1970, around 200 systems for environmental assessment (EA) have been introduced in countries, states and international organisations around the world. EA may be described as an environmental management tool whose objective is to identify, predict and evaluate the potential biological, physical, social and health effects of a proposed development action and to communicate the findings in a way which encourages environmental concerns to be adequately addressed by stakeholders, including decision-makers and communities prior to development decisions being made. It plays a crucial role in environmental protection and meeting the challenges of sustainable development, a view which was recognised within the provisions of the declaration of the United Nations Conference on Environment and Development (UNCED), or the Earth Summit meeting, in Rio de Janeiro, Brazil in 1992 and the recommendations made in the resulting global programme of action (Agenda 21). Principle 17 of the UNCED Declaration states that 'Environmental impact assessment, as a national instrument, shall be undertaken for proposed activities that are likely to have significant adverse impacts on the environment and are subject to a decision of a competent national authority'.

During its development over a period of almost 30 years, the methods and approaches to EA have been tried, tested and refined and, in what is perhaps one of the most important developments, its application has been extended from a project level to an assessment of the environmental implications of policies, plans and programmes (strategic environmental assessment, SEA). Improvements in practice have been achieved, yet it is also recognised that there are many opportunities for strengthening institutional capacity for EA, improving the design of EA systems and their implementation at an operational level to make the process more effective (United Nations Environment Programme 1992; Sadler 1996). In particular, these include: the ways in which issues for inclusion in EA studies are determined (scoping); the analysis of development alternatives; consideration of more complex environmental impacts, especially those which are of a cumulative and/or transboundary nature; improved quality control mechanisms; better public participation in the process; and, as stated above, the further application of EA beyond the project level. Recent and proposed major modifications to established systems for EA, for example in Canada, Hong Kong, New Zealand and the European Union (EU), have responded to some of these needs and recently developed new systems for EA have demonstrated a greater appreciation of their importance (e.g. Namibia, Ghana and Chile).

Terminology

The terms EA and environmental impact assessment (EIA) and environmental statement (ES) and environmental impact statement (EIS) are used respectively to describe the overall process and the written report arising from the studies. Regrettably, there is no real consensus on the use of these terms and they are often used interchangeably. Many countries have EIA systems, whereas the World Bank has procedures for EA. Within the UK, the preferred term has been EA, particularly chosen to avoid the impression that the process is restricted to the analysis of negative impacts, whereas the EU Directive on which the UK system is based consistently uses the expression EIA. The approach taken in this book is to use the term EA for all types of environmental assessment and the term SEA when referring exclusively to the assessment of policies, plans and programmes. When reference is made to the processes established by countries or international organisations, these are referred to by their original formal title, for example the Namibian EA policy or the EIA procedure of Ghana.

General principles of EA

The legal frameworks, procedures and guidelines for EA introduced in countries and organisations around the world follow some generally agreed principles, which have been described as a hierarchy of core values, guiding principles and operational principles (Sadler 1996). These principles are the product of over 25 years' experience with EA, and in the case of operational principles, the inputs of practitioners arising from a series of workshops (for example, Centre for Environmental Management and Planning (CEMP) 1994) organised to feed into the international study of the effectiveness of EA, which took place between 1993 and 1996. These general principles constitute an essential guideline for the EA practitioner.

According to Sadler (1996) the core values of EA are:

- *Integrity*. The EA process should conform to accepted standards and principles of good practice.
- *Utility*. The process should provide balanced, credible information for decision making.
- *Sustainability*. The process should promote environmentally sound development.

The same source gives the main guiding principles as:

- *A well-founded legislative base* with clear purpose, specific requirements and prescribed responsibilities.
- *Appropriate procedural controls* to ensure the level of assessment, scope and consideration and schedules for completion are relevant to the circumstances.

- ◆   *Incentive for public involvement* with structured opportunities tailored to the issues and interests at stake.
- ◆   *Problem- and decision-orientation*, concerned with the issues that matter, the provision of consequential information, and explicit linkage to approvals and condition-setting.
- ◆   *Follow-up and feedback capability*, including compliance and effects monitoring, impact management, and audit and evaluation.

Sadler (1996) also sets out the main operational principles for effective EA practice.

EA should be applied:

- ◆   to all development projects or activities likely to cause potentially significant adverse impacts or add to actual potential foreseeable cumulative effects;
- ◆   as a primary instrument for environmental management to ensure that impacts of development are minimised, avoided or rehabilitated;
- ◆   in a way that the scope of review is consistent with the nature of the project or activity and commensurate with the likely issues and impacts; and
- ◆   on the basis of well defined roles, rules and responsibilities for key actors.

EA should be undertaken:

- ◆   throughout the project cycle, beginning as early as possible in the concept design phase;
- ◆   with clear reference to the requirements for project authorisation and follow-up, including impact management;
- ◆   in accordance with established procedures, best-practice guidance and project-specific terms of reference; and
- ◆   to provide appropriate opportunities for public involvement of communities, groups, and parties directly affected by or with an interest in the project and/or its environmental impacts.

EA should address, wherever necessary or appropriate:

- ◆   other related and relevant factors, including social and health risks and impacts;
- ◆   cumulative and long-term, large scale effects;
- ◆   design, location and technological alternatives to the proposal being assessed; and
- ◆   sustainability considerations, including resources productivity, assimilative capacity and biological diversity.

EA should result in:

- ◆   accurate and appropriate information regarding the nature, likely magnitude and significance of potential effects, risks and consequences of a proposal and alternatives;

- the preparation of an ES that presents this information in a clear understandable manner which is relevant for decision-making; and
- an ES which identifies the confidence limits that can be placed on the predictions made, and which clarifies agreement and disagreement among the parties involved in the process.

EA should provide the basis for:

- environmentally sound decision making in which terms and conditions are clearly specified and enforced;
- the design, planning and construction of acceptable development projects that meet environmental standards and management objectives;
- an appropriate follow-up process with requirements for monitoring, management, audit and evaluation;
- follow-up requirements that are based on the significance of potential effects and on the uncertainties associated with prediction and mitigation; and
- learning from experience with a view to making future improvements to the design of projects or the application of the EA process.

International developments in EA

Prior to EA, development projects were often assessed according to technical, economic and political criteria and the potential environmental, health and social impacts of actions were rarely fully considered. Even where environmental impacts were included, early use of cost benefit analysis (CBA) crudely attempted to place a monetary value upon non-economic variables (e.g. the social and health impact of air pollution; the destruction of marine ecosystems; etc.). As a consequence of such restricted assessment, many developments resulted in unforeseen harmful impacts which reduced their predicted benefits, for example in the case of the Aswan Dam development in Egypt, which created the unpredicted secondary effect of a reduction in the importance of the Mediterranean fishing industry, through a curtailment of the movement of sediment into the Mediterranean Sea.

Growing environmental awareness and increasing public concerns over the impacts of developments on biophysical systems were a major impetus to the US NEPA. The purposes of the Act are 'to promote efforts which will prevent or eliminate damage to the environment and biosphere' (Section 2) using 'a systematic, interdisciplinary approach' which will 'ensure that presently unquantified environmental values may be given appropriate consideration in decision making along with economic and technical considerations' (Section 102).

The implementation of EA systems in other countries following the principles set out in NEPA began in 1973 and 1974 with Canada, Australia and New Zealand. The Philippines was the first developing country to introduce EA (in 1977) and in Europe, systems were first introduced in France (1976) and the Netherlands (1978).

Within the EU, the 1985 Directive on the assessment of the effects of certain public and private projects on the environment (Directive 85/337/EEC) established mechanisms for EA which were implemented as part of individual member state legislation when the Directive came into force in 1988. This process of implementation in the UK was complicated compared with those EA systems developed elsewhere within a framework environmental law in that no single regulation covers all the requirements of the process. A revised version of the European EA Directive was published in March 1997 (European Communities 1997).

EA systems have continued to be introduced in developing countries and countries in transition in Central and Eastern Europe, and more can be expected, particularly given the encouragement of the UNCED declaration (UNEP 1992) and the introduction of procedures for EA by development banks and international development agencies. There are now around 70 developing and transitional countries with EA legislation in place and in some of these countries there is considerable activity with respect to the reorganisation of government responsibilities for EA at a national and regional level, the revision of existing EA systems and the development of more detailed procedures or guidelines to support EA practice. Further details on some of these developments can be found in European Bank for Reconstruction and Development (1994), Wilson *et al.* (1996) and Yeater and Kurokulasuriya (1996).

States and provinces in some federal countries have developed EA procedures. More than 30 US states have established limited forms of EA review or have enacted 'mini-NEPAS' and in Canada, all the provinces and territories have their own EA systems and other systems exist, for example for native land claim settlements (Sadler 1996).

The World Bank introduced procedures in 1989 (amended in 1991) for EA prior to decisions on financing development projects (World Bank 1991) and other development banks such as the Asian Development Bank (Asian Development Bank 1993) have EA procedures in place. International development agencies, such as the Canadian International Development Agency (Canadian International Development Agency undated) have also developed procedures for policy and programme assessment.

The use of EA in business and industry outside the project planning process is growing in importance as organisations seek proactively to manage the environmental consequences of their activities and improve their environmental performance. The Business Charter for Sustainable Development (International Chamber of Commerce 1991) includes as one of its 16 principles of environmental management one calling for an EA to be performed 'before starting a new activity . . . and before decommissioning a facility or leaving a site'. A number of multinational companies have established their own policies and operational guidelines for EA, for example Shell International (Shell International 1994) and further encouragement has come from the World Business Council on Sustainable Development (World Business Council on Sustainable Development 1995), whose business perspective on EA states that it 'can assist companies in their quest for continuous improvement by identifying ways of maximising profits through reducing waste and liabilities, raising productivity and demonstrating a company's sense of duty towards its customers and neighbours'. Many companies seeking actively to demonstrate improvements in environmental performance have, or are in the process of developing and implementing, an environmental management system (EMS) to the requirements of the international standard ISO 14001 (International Organisation for Standardisation 1996). One of the key tasks in the early stages of developing an EMS is the identification of environmental effects and an evaluation of their significance, and this is fundamentally based on the principles of EA. Another example of the use of EA outside national formal procedures is in the formulation of conservation management plans. This can be illustrated with the case of management plans for nature reserves in the UK following the guidelines developed by the Nature Conservancy Council (Nature Conservancy Council 1991).

One of the most significant developments in EA has been the introduction of procedures which address the environmental impacts of policies, plans and programmes, or SEA. In particular, SEA presents a means for evaluating environmental issues at the development policy level rather than waiting to deal with the consequences of these decisions at the project level. SEA also offers a better way of dealing with cumulative impacts, which project-level EA has generally found difficulties with.

The basis for another key development in EA, the assessment of transboundary impacts, exists in the Espoo Convention (Espoo 1991), signed by the European Commission and 29 member countries of the UN Economic Commission for Europe (UNECE) in 1991. The convention would require signatory countries to carry out EAs at an early planning stage for activities likely to cause significant adverse effects of a transboundary nature.

Costs and benefits of EA

EA has a key role to play in the way in which new development proposals are designed, approved and implemented. To be effective, the process needs to be integrated into the project planning cycle, with environmental studies taking place in parallel with the project design as it provides a useful framework within which environmental considerations and design can interact. Used in this way, the EA can indicate how the project design might be modified to anticipate and minimise possibly adverse effects and provide for a better environmental option or alternative process/design/location (Department of the Environment UK 1989).

Some of the main criticisms voiced against EAs, particularly by developers, are that they are expensive to implement, notably in areas where little is known about existing environmental and social conditions, take up a lot of time and generally create additional bureaucracy to deal with. Design changes produced as a result of EA findings may also increase capital costs. In most EA systems, the cost is borne by the proponent of the development, although it is quite difficult to obtain information on these costs as this is normally regarded as confidential information by the consultants commissioned to make the studies. The costs of conducting EAs are dealt with further in Chapter 6.

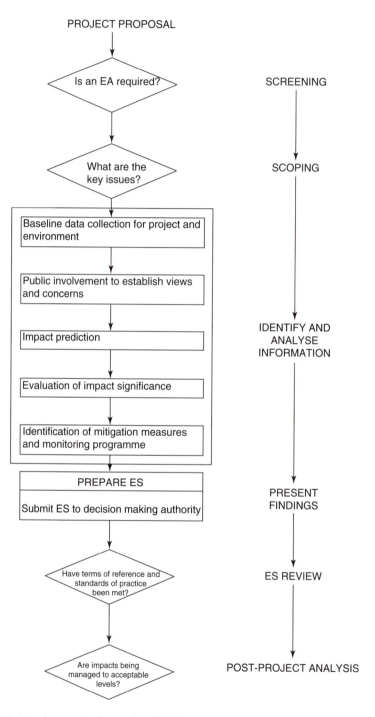

FIGURE 1.1 A generalised procedure for EA

The use of EA may, however, reduce long-term project costs, avoid non-compliance with laws and regulations and reduce the time taken to reach a decision by identifying and quantifying the beneficial effects and those environmental consequences of the development which might require expensive pollution clean-up and abatement technology, compensation payments or other costs at a later date. Effective EA may also benefit the environment, for example by lowering health and social costs, and helping to protect environmental goods and services and to maintain biodiversity. Many of the environmental amenities that would otherwise have been degraded or destroyed have a unique value, which over time will far outweigh EA costs. Many cases show that the use of EA has allowed the choice of an option which is both environmentally and economically superior to the original choice.

Overview of the main stages in the EA process

While the details and relative importance of components of EA procedures differ among countries and organisations, there exist a common series of stages for project-level EA (Figure 1.1). These begin with a determination of the need for EA (screening), followed by an initial analysis of the proposal to establish what are the main issues for inclusion in the study (scoping). More detailed assessment then involves the collection and analysis of information (baseline description and impact prediction and evaluation) and the views and concerns of stakeholders (public involvement), leading to the production of a report (ES) which describes the nature of the project (and possibly a range of alternatives), its environmental setting, the impacts associated with the development and proposals for dealing with those impacts considered to be potentially significantly adverse (impact mitigation). On presentation of the ES (draft or final version) to the decision-making authority, there is a review of the statement to check that terms of reference for the project and standards of acceptable practice have been met (ES review). Finally, there may be post-project activities in which impacts and environmental management plans are monitored and audited.

In the following section, the main stages of the EA process are briefly described. Further consideration of the specific aspects of EA practice is given in the following chapters: impact identification, prediction and evaluation (Chapters 2 and 3); public involvement (Chapter 5); ES review and post-project analysis (Chapter 7).

Although this general framework for EA is presented as a series of stages, in practice EA is an iterative process in which, for example, discussions with stakeholders take place during scoping and at other points to help refine impact evaluations; and further baseline studies may be required later in the process if it is found there are insufficient data adequately to predict an impact. Overlying all of these stages is the need to manage the process with respect to personnel involved, budgeting and liaison with client(s), authorities and other stakeholders (see Chapter 6).

Is an EA required? (screening)

The first stage in the EA process is to determine whether or not a proposed development activity requires to be subjected to an EA. The term screening is commonly used to describe this process, which should be distinguished from the next stage of scoping, a term which is used here to describe the process by which the range of issues to be included in an EA study are decided upon.

A number of screening methods have been devised. These include the use of positive and negative lists, screening matrices and initial environmental evaluation (IEE) (see Chapter 2). Whatever the methods used, screening needs to be relatively quick (to avoid delay to projects not requiring EA) and easy to use, yet be sufficiently comprehensive to ensure that all projects that warrant EA are clearly identified. Screening often utilises criteria and thresholds related to the project (which reflect the potential of the project to give rise to adverse effects) and to the environment (which reflect the ability of the location to accommodate the proposed project). The most effective screening procedures include both project and environment criteria/ thresholds.

A commonly adopted approach to screening has been to produce lists of projects requiring an EA (positive screening lists). Examples include procedures in operation in Malaysia, Thailand and the EU countries. Within Europe, EC Directive 85/337 includes two annexes listing projects having either a mandatory (Annex I) or discretionary (Annex II) requirement for EA. The criteria for including projects in either of these lists are based on the scale and size of the proposal, the nature of the activities and the sensitivity of the environmental setting. Accompanying thresholds help to define more closely which projects require EA. For example, thermal power stations have a mandatory requirement for EA only where they exceed 300 MW. Another example, from the UK Regulations which implemented the EC Directive in 1988, states that new road schemes may require an EA (Annex II project) if their length exceeds 1 km and their route passes through a National Park or through or within 100 m of a Site of Special Scientific Interest (SSSI), National Nature Reserve (NNR) or a conservation area.

A negative screening list is the opposite of the above in that EA is required for a project unless it is included in the list. This categorical exclusion procedure is a more difficult system to operate due to the fact that lists of excluded projects can be very large unless regulations are skilfully drafted.

Although decisions on whether or not EA is needed may be quick and easy to make for certain high-profile projects, there are many borderline cases where this is more difficult. Several countries, including Canada and Thailand, have developed a procedure which more carefully examines the need for EA by conducting an IEE on a project prior to any requirement for a full-blown assessment. This two–stage screening process (Figure 1.2) has the benefit of using resources more efficiently. The results of the IEE may be sufficient to grant planning permission for a project which might otherwise have been delayed by an (unnecessary) extensive EA. Where a more detailed study is required, the findings of the IEE can be fed into the final statement, thus avoiding any additional costs.

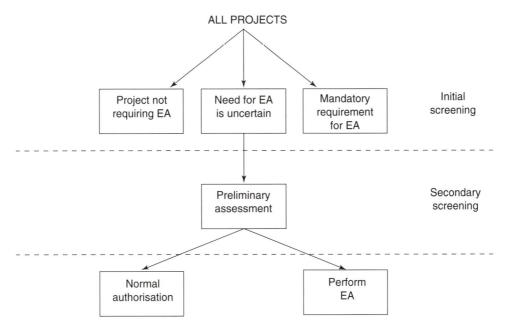

FIGURE 1.2   Two-stage screening procedure

What are the main issues? (scoping)

Scoping refers to the process of identifying, from a broad range of potential problems, those key issues that should be addressed by an EA. The importance attached to scoping arises from the fact that EAs are usually conducted under serious time, budgetary and resource limitations. Any priority-setting activity, therefore, should improve efficiency and provide a more focused product (the ES) for decision-makers. During the early stages of using EA in the USA, many ESs were encyclopaedic, included too much irrelevant information and were a burden to the decision-making process. Many methods have been developed to assist scoping. Depending upon the nature of the priority issues identified, the baseline study can be structured around the results of the scoping exercise.

Collection and analysis of information

This stage of EA involves a number of linked steps in which the baseline environmental conditions and characteristics of the development are described and impacts associated with a project, or series of project alternatives, are identified, their potential magnitude predicted and significance determined. This stage is also about determining the views and concerns of the public and agencies/organisations (governmental and non-governmental) who are known to have or who may have an interest in the environmental and social consequences of a development proposal (collectively described here as stakeholders). On the basis of these studies, recommendations on how to prevent or mitigate potentially significant adverse impacts (and highlight the environmental benefits) of the proposal can be developed.

A wide range of methods, including checklists, matrices, networks and overlays, have been developed to identify the potential impacts of a development proposal. As well as these simple methods, more quantitative approaches are used in which impact scores for a number of alternatives are derived. Methods for impact identification, prediction and evaluation are considered in Chapters 2 and 3.

Public involvement

Public involvement should be an integral part of any EA system. Efforts should be made to obtain the views of, and to inform, the public and other interest groups that may be directly or indirectly affected by the project. The authorising agencies may not always identify the environmental issues which the public perceives to be important and they may also lack the detailed local knowledge that the public possesses. Advantages of participation may lead to the provision of information about local environmental, economic and social systems; the possible identification of alternative actions; an increase in the acceptability of the project as the public will better understand the reasons for the project; and a minimisation of conflict and delay. Problems may nevertheless arise. Public participation may, in the short term, be time-consuming and increase costs, and participants may be unrepresentative of the community. In spite of these potential problems, many countries are actively encouraging public involvement in EA and the World Bank has now made it a legal requirement before any loan can be made. Public consultation and participation are considered further in Chapter 5.

Communicating the findings

The principal objective of EA is to provide decision-makers with an account of the implications of proposed courses of action before a decision is made. The results of the assessment are assembled into a document often referred to as an ES, although many other terms have been and still are used. The ES contains a discussion of beneficial and adverse impacts considered to be relevant to the development action being investigated and is one component of the information upon which the decision-maker ultimately makes a choice. At this stage, other factors such as unemployment, energy requirements or national policies may influence the outcome of the decision. A final decision can be made with due regard being paid to the likely consequences of adopting a particular course of action, and where necessary by introducing appropriate monitoring programmes. The principal elements of project management and ES report presentation are given in Chapter 6.

ES review

Because most EA systems state that an assessment must be produced by the project proponent, there is usually a need for an impartial, scientific and independent review to ensure that an ES is sufficiently objective and impartial and that it covers all pertinent issues and conforms to procedural requirements in place. The review is often likely to be directed by the authority from which authorisation for the development is requested and may involve the use of an independent panel of experts. ES review is considered further in Chapter 7.

Post-project analysis

From a quality assurance perspective, it is important that the EA procedure continues after the decision-making stage.

The use of monitoring and auditing in EA allows the process to be more interactive and to provide checks on mitigation measures and to verify the accuracy of predictions made. Such post-project analysis studies (which are considered further in Chapter 7) also help with management of the existing project and to improve the design of future projects and their assessment.

Questions for thought

1. How can EA contribute to more environmentally sustainable development?
2. What are the benefits of performing EA?
3. How can the EA process contribute to the development of environmental management systems?
4. Noting the component parts of an EA system, develop and refine your own EA procedure.

C h a p t e r   2

Environmental assessment methods

◆ **Introduction**   16
◆ **Baseline studies**   18
◆ **EA identification methods**   18
◆ **Questions for thought**   32

Introduction

In this chapter, the nature and characteristics of impacts and methods used in their identification such as checklists, matrices, networks and environmental features mapping are discussed. These may be used during a scoping study to help determine which are the priority issues, or later in the EA process as an aid to determining the requirements for impact prediction and mitigation. Predictive techniques are considered in more detail in Chapter 3.

An environmental impact is an event or effect, which results from a prior event. It can have both spatial and temporal components and can be described as the change in an environmental parameter, over a specific period and within a defined area, resulting from a particular activity compared with the situation which would have occurred had the activity not been initiated (Wathern 1989). The impact is the difference between the with-project and without-project condition, which may be possible to quantify (for example a predicted change in an environmental parameter such as a noise level). Alternatively, the impact prediction may be made more subjectively through literature review and value judgement (for example a change in landscape value) (see Chapter 3).

Different impacts are likely to arise at different times during the life of a project, and so a phased approach to EA is a good practice to adopt. The main phases of a project (not all of which are relevant to all projects) are: pre-construction and planning, construction, operational and post-operational (or decommissioning). Consider for example a new rail development. Uncertainty about the preferred route can cause numerous socio-economic impacts during the pre-construction phase, arising from changes in house prices (increase or decrease) and the movement of businesses. Increased intensity of traffic during construction may well have detrimental socio-economic impacts, some of which become beneficial once the new route has been opened. Post-operational impacts may then arise from a subsequent change to the completed transport system such as the closure of a commuter railway station, leading to traffic impacts as commuters resort to travelling to work by car. Some impacts present themselves immediately (for example constructional noise causing disturbance to local residents), while others appear only after a protracted period (for example the bio-accumulation of heavy metals by terrestrial and aquatic organisms).

Environmental systems are composed of complex inter-relationships of linked individual components and sub-systems. Consequently, impacts on one component may well have effects on other components, some of which may be spatially (and temporally) distant from the component immediately affected. Indirect impacts (also called secondary, higher-order or knock-on impacts) may be difficult to identify, evaluate and predict, but should not be excluded from impact studies (see Figure 2.1).

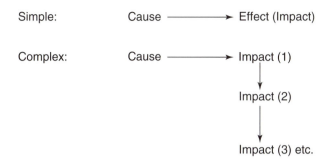

FIGURE 2.1 Direct and indirect impacts

While some impacts are irreversible (e.g. the loss of ancient woodland during road construction), others are reversible (e.g. noise levels during construction). Given that the responses of ecosystems to externally induced changes are not always known, there may be difficulty with precision when assessing environmental impacts. In many cases, the likelihood or probability of an impact occurring is uncertain and is described generally (a hierarchy of certainty/uncertainty terms is common in ESs, for example unlikely, possibly, likely, probably, almost certainly) rather than in any quantified manner. To date, the use of statistical probability has been applied mainly to risk assessment (see Chapter 4).

Cumulative impacts of developments take a variety of forms such as: frequent and repetitive or high density impacts on a single environmental medium; synergistic effects from multiple sources on a single environmental medium; impacts resulting some distance from the source; and secondary impacts resulting from a primary activity (Department of the Environment UK 1995). Cumulative effects assessment (CEA), which attempts to analyse and predict the potential for a range of effects accumulating from actions over time and space, is best performed at a programme or policy level rather than at the project level (see Chapter 8).

Not all individuals or social groups will agree on the environmental changes that should be classed as impacts for inclusion in an ES. The proposed siting of a new supermarket complex on land with local nature conservation value might be welcomed (e.g. beneficial impact) by an unemployed retail worker but considered as a detrimental impact by a member of the local natural history group. Seeking the views of the local community by means of scoping is an important early stage in the EA process.

According to the EC Directive (European Council 1985) and UK Regulations (Department of the Environment UK 1988), EA should be performed for those projects 'which are likely to have significant effects on the environment by virtue of their nature, size or location'. Determining impact significance will involve discussions with relevant organisations, experts and members of the public, and where it is based on value judgement, is often the subject of considerable debate. Impact significance should be distinguished from impact magnitude, which can be determined by means of some observation or experiment.

Baseline studies

Baseline studies consist of a description of those aspects of the physical, biological and social environments which could be affected by a proposed development. These are fundamental to technical assessment studies to enable the level of significance of impact to be determined in relation to existing baseline conditions (see Chapter 3). They need to be conducted early on in the EA process, usually following scoping, since they provide information on the 'before-project' conditions, which need to be established before the identification and prediction of impacts can be made. Baseline studies, however, may not take place exclusively at this early stage. Additional baseline data may be required later on, for example to help refine impact predictions. Baseline studies can account for a large part of the overall cost of an EA, particularly where they require extensive field studies. Although new data may need to be acquired, existing information from appropriate sources should be regarded as a valuable and vital resource to be used whenever possible.

EA identification methods

EA methods may generally be regarded as mechanisms by which information is collected and organised, evaluated and presented. There may be provision to assess the magnitude and significance of impacts, and this may be quantified, particularly when a number of alternatives are being evaluated. It is usual for large data sets to be gathered, and careful consideration should be given to how this can effectively be communicated to decision-makers, consultees and members of the public, not all of whom will be technical experts. A distinction is often made between methods and techniques (see Chapter 3): methods are normally concerned with impact identification and may include guidance on impact evaluation; techniques are concerned with predicting future states of environmental parameters, such as air and water quality and may involve mathematical modelling. While there may be a formal obligation to make reference to their use in the ES in some EA systems around the world, in Britain their use has tended to be internal, i.e. they are used by EA practitioners but not always included in the statement.

There are many types of methods available to the EA practitioner, Canter (1992) and Bisset (1992) provide a thorough review and description of them. It is essential that the user is not overawed by them and does not become lost in their complexity. It is important to remember that methods are a tool within the EA process to scope potential impacts and process information. They are not the panacea for all EA problems. A great deal of research effort has gone into their development over the past two decades. The fundamental question that must be asked, however, is whether this effort has significantly influenced the evolution of EA practice. The results of a comprehensive analysis of the use of EA by Caldwell *et al.* (1982) seem to condemn the work of many academics to the category 'irrelevant to real needs'. About 30 per cent of the EAs reviewed were prepared using *ad hoc* methods generally devised by the study team specific to their needs.

A family of checklist methods are available offering a range of characteristics and complexity. They are commonly used to solve the problem of what aspects of a development to consider in the assessment. Checklists provide an *aide-mémoire* for project managers/practitioners. They are designed to: ensure that all potential impact areas are considered; provide a structured approach for identifying key impacts and/or pertinent environmental factors for consideration in environmental impact studies; and stimulate team discussions during the planning, execution, and summarisation of the EA. Checklists can be readily modified to make them more pertinent for different project types in given locations. Like all methods, checklists are an iterative process, on-going, and are likely to be continually refined throughout the EA process as more project information becomes available to the practitioner or as he or she becomes more familiar with the project.

Basic checklists are simply a list of those physical, biological and socio-economic factors which may be affected by a development, with no attempt to evaluate impacts qualitatively or quantitatively. Guidelines on EA procedures for the UK include a checklist of matters which may need to be included in an ES (Department of the Environment UK 1989). Their drawback is that they do not take account of indirect impacts nor show cause–effect relationships between project activities and environmental attributes (Figure 2.1). The issues of likely concern relevant to proposed landfill sites using a checklist format are:

◆   nuisance (odour, dust, litter, birds, vermin, noise, etc.);
◆   traffic;
◆   landscape and visual impact;
◆   water (ground and surface);
◆   landfill gas (fugitive, controlled (flaring, etc.));
◆   ecology.

For a descriptive checklist the basic checklist is extended to include information on data requirements, sources of information and predictive techniques likely to be used, although the relative importance of different impacts is not determined (Table 2.1).

Other checklist methods include scaling checklists which rank listed impacts in order of magnitude or severity. These may be scored on an interval or ratio scale and in some cases aggregated, for example when project alternatives are being compared. Although scaling checklists offer some interpretation of impacts, they tend to rely upon the subjective assignment of numerical values, which if simply aggregated (by arithmetic addition) would imply that each impact has the same importance. This may well be misleading and so the method should be used with some caution. Weight-scaling checklists have been developed as a means of evaluating impacts. Paired comparison checklists are a method of evaluating each pair of alternatives for a project by creating checklists of scoped environmental issues, which may be scaled.

TABLE 2.1 Example of information included in a descriptive checklist

Data required	Information sources/predictive techniques
Health	
Change in air pollution concentrations by frequency of occurrence and number of people at risk.	Current ambient concentrations, current and expected emissions, dispersion models, population maps.
Nuisance	
Change in occurrence of visual (smoke, haze) or olfactory (odour) air quality nuisances, and number of people affected.	Baseline citizen survey, expected industrial processes, traffic volumes.
Water quality	
Changes in permissible or tolerable water uses and number of people affected – for each relevant body of water.	Current and expected effluents, current ambient concentrations, water quality model.
Noise	
Change in noise level and frequency of occurrence, and number of people bothered.	Changes in nearby traffic or other noise sources, and in noise barriers; noise propagation model or nomographs relating noise level to traffic, barriers, etc.; baseline citizen survey of current satisfaction with noise levels.

Source: Schaenam (1976)

Some checklists have been produced by developers/agencies for specific project types. An example is the checklist produced by the National Rivers Authority (NRA) in the UK for land drainage improvement works (Thames Region NRA 1989). Such developments are performed under the authority of the NRA and may be subject to EA. In order to help in compliance with these regulations, Thames Region NRA produced a series of EA guidelines. These include a screening procedure and a comprehensive checklist of environmental factors that may need to be covered in the assessment. These are considered in 15 sections under the following headings: physical characteristics of the site and its surroundings; ecological characteristics; human activity patterns in the area; infrastructure services and existing pollution levels.

Matrix methods

Viewed simply, an EA matrix consists of two checklists: one is a set of development project actions, the other is a set of environmental components. Matrices can be designed to identify impacts associated with the various phases of a development and specific environmental systems/sub-systems if this is desired. Figure 2.2 shows a matrix used for a proposed power plant development. The boxes or cells which make up the matrix can be scored to provide an easily understood summary of where interactions between project and environment (the impacts) arise.

Perhaps the best known application of methods is the Leopold matrix (Leopold *et al.* 1971) which is an example of a presentational matrix in which impact magnitude and significance (importance) are recorded within each impact cell, using a scale of 1-10 (10 is greatest). Numerical ranges other than 1-10 (e.g. negative values to denote negative impacts) and various descriptive and symbolised matrices have been used (e.g. distinguishing between direct and indirect impacts) (Shopley and Fuggle 1984). The matrix was designed as a comprehensive method for all potential activities of the US Geological Survey and is consequently relatively large and unwieldy. There is wide acceptance that it is a general listed matrix and that many of the potential interactions would not be applicable to a particular situation. In fact, it was estimated that only approximately 20–50 potential interactions were likely to be significant in most proposals.

A criticism similar to that made of checklists has been raised for matrices, that is, they compartmentalise the environment into separate items identified by a series of discrete two-way linkages between development activities and components (Canter 1992 and Bisset 1992). They tend to concentrate on primary impacts, with less attention paid to secondary impacts. Comparisons between project alternatives may be difficult (unless weight-scaled impact scores are used) and replication of the method could be difficult, given the subjective judgements used in scoring impact significance. Figure 2.3 shows a matrix used for comparing alternative project sites.

Component interaction matrices have been designed for identifying indirect impacts. Originally proposed by Environment Canada (1974), the method has been further refined by the use of a quantitative adaptation (Wathern 1984). This approach relies upon the use of matrix algebra to identify indirect, second- and higher-order impacts from an analysis of direct effects alone.

Another variation to the application of matrices is the two-stage screening procedure adopted in Canada. The initial, so-called Level 1 matrix is constructed as described earlier and scored to denote where impacts arise at the four stages associated with the implementation of the project, namely: site investigation and preparation; construction; operation and maintenance of the completed project; future and related activities subsequent to project development. Each of these potential interactions is subject to more detailed scrutiny in the Level 2 matrix which aims to achieve the following: distinguish between significant and insignificant impacts; identify activities for which a design solution or mitigating measure is available; identify activities which have unknown or potentially adverse consequences. This may involve consultation with experts, the initiation of a preliminary data collection scheme and discussions with the proponent and local people. In the light of these deliberations, the consequences of each potential interaction can be assessed. Screening decisions are influenced by a number of considerations (Federal Environmental Assessment Review Office 1986). These are:

KEY:

✕ Negligible or no effect
○ Potential negative effect
+ Potential positive effect
✳ Potential positive and negative effects

Effects on:				Development Activity	A SITE PREPARATION AND CONSTRUCTION	1 Traffic Generation	2 Improvement of Access Road	3 Site Levelling
PHYSICAL/CHEMICAL EFFECTS	Water	A	Groundwater					
			Flow & Water Table Alteration			✕	✕	✕
			Interaction with Surface Drainage			✕	✕	✕
			Water Quality Changes			✕	✕	✕
		B	Surface Water					
			Drainage Characteristics			✕	+	✳
			Flow Variation			✕	✕	✕
			Water Quality Changes			✕	✕	✕
	Land		Soil Quality			✕	✕	✕
			Soil Structure			✕	✕	✕
			Compatibility of Land Uses			✕	✕	✕
			Compaction & Settling			✕	✕	○
			Stability			✕	✕	○
			Landscape Character			✕	✕	○
			Geological Resources			✕	✕	✕
	Atmosphere		Air Characteristics			○	✕	✕
			Wind			✕	✕	✕
			Micro-climate Changes			✕	✕	✕
			Macro-climate Changes			✕	✕	✕
ECOLOGICAL EFFECTS	Ecology	A	Terrestrial					
			Flora			✕	✕	+
			Fauna			✕	✕	+
		B	Aquatic					
			Flora			✕	✕	○
			Fauna			✕	✕	○
HUMAN EFFECTS	Humans	A	Nuisance					
			Noise/Vibration			○	○	○
			Litter/Debris/Dust			○	○	○
			Odour			✕	✕	✕
			Pests/Vermin			✕	✕	✕
		B	Visual/Recreational Amenity					
			Landscape Modification			✕	✕	✳
			Visual Obtrusion			✕	✕	✕
			New Landscape Feature			✕	✕	✳
		C	Health and Safety					
			Health			○	○	○
			Safety			○	○	○
		D	Socio-economics					
			Social Welfare			✕	✕	✕
			Economic Welfare (jobs)			+	+	+
TRANSPORT EFFECTS	Transport		Road Capacity			○	✕	✕
			Road Safety			○	✳	✕
			Highway Infrastructure			○	✕	✕
CULTURAL HERITAGE EFFECTS	Cultural Heritage		Sites of Archaeological Interest			✕	✕	✕
			Ancient Monuments			✕	✕	✕
			Listed Buildings			✕	✕	✕

FIGURE 2.2 Matrix of a proposed power plant
Source: Harrop (1994), reprinted with permission of The Royal Society of Chemistry

4 Drilling and Piling	5 Design and Plant	6 Installation of Utilities	B TRANSPORTATION OF SLUDGE	1 Transport by Road	2 Spillage/Leaks,Accidents	C COMMISSIONING MAINTENANCE AND OPERATION	1 Stack Emissions	2 Liquid Emissions	3 Equipment Operation	4 Operational Failure	5 Storage of Ash	D DISPOSAL OF ASH	1 Transport of Ash	2 Disposal of Ash to Landfill	E DECOMMISSIONING	1 Demolition Activities	2 Reclamation
X	X	X		X	X		X	X	X	X	X		X	X		X	X
X	X	X		X	X		X	X	X	X	X		X	X		X	X
X	X	X		X	O		X	O	X	O	O		X	O		X	X
X	X	X		X	X		X	X	X	X	X		X	X		X	+
X	X	X		X	X		X	X	X	X	X		X	X		X	X
X	X	X		X	O		X	O	X	O	O		X	O		X	X
X	X	X		X	X		X	X	X	X	X		X	X		X	X
X	X	O		X	X		X	X	X	X	X		X	X		X	X
X	*	X		X	X		X	X	X	X	X		X	X		X	+
X	X	X		X	X		X	X	X	X	X		X	X		X	X
X	X	X		X	X		X	X	X	O	X		X	X		X	X
X	*	X		X	X		X	X	X	X	O		X	X		*	+
X	X	X		X	X		X	X	X	X	X		X	X		X	X
X	X	X		O	X		O	X	X	O	X		O	O		X	+
X	X	X		X	X		X	X	X	X	X		X	X		X	X
X	X	X		X	X		O	X	X	O	X		X	X		X	X
X	X	X		X	X		O	X	X	O	X		X	O		X	X
X	+	X		X	O		O	X	X	O	X		X	X		O	+
X	*	X		X	O		O	X	X	O	X		X	X		O	+
X	X	X		X	O		O	O	X	O	X		X	X		O	+
X	X	X		X	O		O	O	X	O	X		X	X		O	+
O	*	O		O	X		X	X	O	O	X		O	X		O	X
O	X	O		O	O		X	X	X	O	O		O	X		O	X
X	X	X		X	O		O	X	X	O	X		X	X		X	X
X	X	X		X	X		X	X	X	X	X		X	X		X	X
X	*	X		X	X		X	X	X	X	O		X	X		O	+
X	*	O		X	X		O	X	X	O	O		X	X		*	X
X	*	O		X	X		X	X	X	X	O		X	X		*	X
O	*	O		O	O		O	O	O	O	X		O	O		O	X
O	*	O		O	O		O	O	O	O	X		O	O		O	X
X	+	X		X	X		X	X	X	O	X		X	X		X	+
+	X	+		X	X		X	X	+	O	X		X	X		+	+
X	X	X		O	O		X	X	X	X	X		O	X		X	X
X	X	X		O	O		X	X	X	O	X		O	X		X	X
X	X	X		O	X		X	X	X	O	X		O	X		X	X
X	X	X		X	X		X	X	X	X	X		X	X		X	X
X	X	X		X	X		X	X	X	X	X		X	X		X	X
X	X	X		X	X		X	X	X	X	X		X	X		X	X

23

			Sites:	Site 1	Site 2	Site 3	Site 4
Phase 1	Physical	Geology					
		Hydrogeology					
		Hydrology					
		Climatology					
		Slope Stability					
		Flooding					
		Atmospheric Dispersion					
		Surface Water Recharge					
		Natural Hazards					
	Political						
	Technical	Groundwater Recharge					
		Land Take					
		Unstable Ground					
		Geotechnics					
		Subsidence					
	Economic	Maintenance					
		Investment					
	Legislative	Planning					
		Environmental					
	Infrastructure	Roads					
		Services					
		Fuel					
Phase 2	Water	Groundwater					
		Surface Water					
	Land						
	Atmosphere						
	Ecology	Terrestrial					
		Aquatic					
	Humans	Nuisance					
		Visual/Recreational Amenity					
		Health and Safety					
		Socio-economics					
	Transport						
	Social	Religion					
		Cultural Heritage					
Phase 3	Buffer Zones						
	Cumulative Impacts						

KEY:
X   Negligible or no effect
0   Potential negative effect
+   Potential positive effect
*   Potential positive and negative effects

FIGURE 2.3   Matrix for the determination of alternative project sites

◆ magnitude – the potential severity of each potential impact and whether it is reversible. In the case of reversible impacts, the rate of recovery or adaptation is important;
◆ prevalence – the likely frequency of similar activities and their potential cumulative impact;
◆ duration and frequency – whether an activity is continuous or intermittent and if intermittent whether there is adequate opportunity for recovery between events;
◆ risk – the probability of a serious environmental effect occurring;
◆ importance – the significance and value attached to a particular resource at present; and
◆ mitigation – whether the potential impacts that have been identified can be resolved with the application of current technology.

After each potential impact has been considered, it is necessary to arrive at a view concerning the aggregate consequences of all of the potential impacts in order to determine whether an EA is required. Clearly, three likely outcomes are possible. First, the scoping study may reveal that an EA is not required. Secondly, impacts may be so significant as to warrant preparation of an EA. Finally, considerable uncertainty may remain. In this last situation an EA may be prudent. Alternatively, the decision may be postponed until a more detailed appraisal has been carried out.

A series of generic steps should be followed when preparing a simple interaction matrix, for example (Canter 1996):

1. list all anticipated project actions and group in temporal phases (i.e. construction, operation, post-operational phases);
2. list pertinent environmental factors from environmental setting and group according to physical/chemical, biological, cultural, socio-economic, and spatial considerations;
3. discuss preliminary matrix;
4. decide on impact 'rating' scheme (numbers, letters, or colours, etc.);
5. score the matrix; and
6. produce a key and consider the inclusion of accompanying notes to explain the matrix.

Networks

Network methods illustrate the (often) multiple impact linkages between project actions and environmental components, including any intermediary links. They are therefore a useful way of presenting indirect and direct impacts together and can assist in preparing specific recommendations for impact mitigation. Networks can also show cumulative and synergistic effects. The main drawback to network methods is that they can be time consuming to construct and may get visually complicated. For this reason it is often more beneficial to create networks for specific environmental

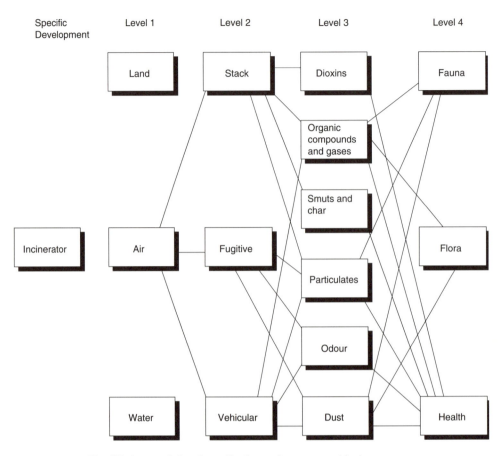

Specific Development	Level 1	Level 2	Level 3	Level 4
	Land	Stack	Dioxins	Fauna
			Organic compounds and gases	
			Smuts and char	
Incinerator	Air	Fugitive		Flora
			Particulates	
			Odour	
	Water	Vehicular	Dust	Health

FIGURE 2.4   Simplified network for air-quality issues for a proposed incinerator
Source: Harrop (1994), reprinted with permission of The Royal Society of Chemistry

systems/sub-systems. Some computerised methods have been developed, for example the IMPACT network developed for forest developments in the United States (Thor *et al.* 1978). Figure 2.4 shows an application of a simplified network for a proposed incinerator (Harrop 1994).

Environmental features mapping

Many of the impacts of development projects have a spatial component and can be most readily identified and assessed through the use of map data and the production of maps showing the extent of impacts likely to occur with development. Environmental features mapping (also known as an overlay or cartographic method) was originally developed by McHarg (McHarg 1968 and 1969) to consider the broad environmental implications of the selection of highway routes. Overlay maps representing the spatial

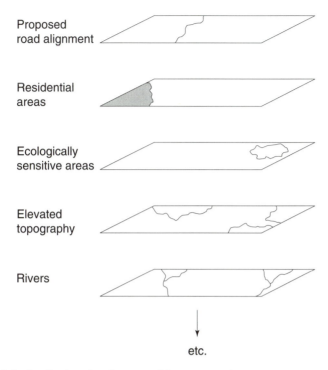

Proposed
road alignment

Residential
areas

Ecologically
sensitive areas

Elevated
topography

Rivers

etc.

FIGURE 2.5   Application of environmental features mapping

variation of an environmental parameter or set are produced on transparent acetate sheets. The degree of impact can be shown by the degree of shading (for example dark tones denote serious impacts; lighter grey tones less serious; no shading denotes no impact) or by colour coding. Overlays can be combined in a variety of ways to show either total impact or impact on selected aspects, for example on ecological or social impacts. Figure 2.5 shows the application of the methodology for a proposed linear scheme (e.g. road, rail, pipelines, waterways, etc.).

There are a number of benefits in this approach which lends itself especially to EA of linear developments: the results presented as a diagram are easy to understand; the spatial distribution of beneficial and adverse impacts is shown; and this can be related to human and natural populations inhabiting the areas affected. Although conceptually a simple method, there are practical difficulties in the manual application of overlays. An important constraint is the limited number of overlays that can be considered at any instant. Interpretation of more than a dozen overlays is often difficult and the results confused. Steinitz, Parker and Jordan (1976) claim that the aggregate map must be recoded and redrawn before it is of much value for analysing impacts, and this greatly increases the inefficiency of the approach. The inflexibility and inefficiency created by the small number of parameters that can be included in any overlay analysis using a manual approach are likely to prove an important constraint when dealing with large and complex development proposals. This problem can, however, be overcome by aggregating associated parameters (e.g. soil properties such as nutrient status, porosity and stoniness may be aggregated to provide a parameter indicative of soil quality).

Many of the constraints of the manual overlay approach are removed by the use of Geographical Information Systems (GIS). In GIS a computer file is produced which contains the digitised data for each parameter. These data can be accessed independently and analysed in any combination for a particular proposal, the original data matrix remaining unmodified. Effectively, an unlimited number of parameters can be considered. A data file is prepared by subdividing the area under consideration into a number of grid squares and recording data from each square. The utility of the computer-based approach lies in the ability to process data rapidly, assess parameters in a variety of combinations from a common data set and assess a range of alternatives rapidly. This allows various facets of the projected development impact to be appraised independently and with different weighting schemes. Aggregate impacts can be represented as a numerical or shade intensity map in the computer print-out.

CASE STUDY: RIVER DON CROSSING, ABERDEEN

Table 2.2 shows a matrix used in a scoping study of the proposed third River Don crossing in Aberdeen, Scotland which considers direct and indirect impacts during the bridge construction and operational phases (AURIS Environmental 1992).

The need for an additional crossing of the River Don in Aberdeen to alleviate congestion over existing crossings was recognised by Grampian Regional Council during the 1970s and led to the production of a scoping and screening study in 1992. It was believed at the time that only a third river crossing could meet likely future traffic flows arising from housing developments north of the city, helping to alleviate demand on existing crossings and generally to reduce congestion on the road networks.

The proposed new bridge (box girder design in concrete or steel) would consist of a three-span structure, about 15 m above the level of the river, with a pier on each bank. The central section would require a span of 65–100 m and an overall length of 125–170 m, depending on the precise alignment. A new section of road (1,500 m) would be mainly at ground level, although given the markedly different bank heights on either side of the river, the northern section would need to have been built up on to an embankment as it approached the river.

The River Don estuary lies about 2 km downstream from the site. The development site itself comprised a long isthmus of land, bounded on three sides by a sharp meander of the river. The isthmus formed a plateau, 15–20 m above the river, with steep wooded sides down to the river bank. The land was formerly worked for agriculture but has now reverted to scrub and rank grassland. At the lower, south-eastern corner is Kettock's Mill (now disused).

The scoping study considered land use, ecological, archaeological and historical features, local populations, recreational uses of the area, noise and traffic flows. A number of organisations were consulted to ascertain their views on the range of

TABLE 2.2 Scoping matrix used for the proposed third River Don crossing, Aberdeen

Impact type	Direct impacts		Indirect impacts	
	Construction	Operation	Construction	Operation
River bank erosion[a]	0(–1?)	–1?	0	–1?
Fishery effects[b]	–1	0	0	0
Noise and vibration[c]	–1	–1	0	0
Air quality[d]	–1	–1?	0	+1?
Visual impact[e]	–1	–1	0	0
Historic and archaeological features[f]	–1	–1	0	0
Nature conservation[g]	–1?	–1?	0	0
Accidents[h]	?	?	0	0
Severance/recreation[j]	–1	–1	0	+1

Key: 0 Unlikely to be a significant impact
      1 Likely to be a significant positive (+) or negative (-) impact
      ? Insufficient information to make a judgement

Notes:

a Depending on results of more detailed consideration of changing patterns of river flow arising from embankment effects: data available. Low possibility of increased bank erosion arising from increased recreational use of the riverside area.

b Siltation and/or disturbance effects of construction standard: operating procedures for mitigation.

c Construction plan operating on Kettock's Mill likely to be very intrusive to local residents. Traffic flow on new access road and local network likely to generate noise levels greater than present. Data on current noise levels needed.

d Dust during construction and increased CO and NO_x from motor traffic in operation. Reduced air pollution levels possible at other locations resulting from diverted traffic using new crossing. Predictions required based on modelled projections of traffic flow.

e Bridge visible from riverside recreational area in wooded valley. Presence of existing paper mill already degrades visual quality of area. Access road to bridge presents visual impacts on Kettock's Mill.

f Possible homestead moat on route of access road. Archaeological assessment required.

g No significant intrinsic conservation value associated with site. Value lies in the integration of the site with the rest of the river valley.

h No information on the projected level of accidents arising from the development has been sought. Estimates based on projected changes in traffic patterns should be made.

j Construction phase will limit recreational access to site. In operation, recreation on site compromised by presence of road, but possible pedestrian access across river from south (indirect effect) could enhance recreational enjoyment of lower Don Valley.

potential environmental impacts. These included statutory organisations responsible for river quality (e.g. North East Rivers Purification Board, now part of the Scottish Environment Protection Agency (SEPA)) and nature conservation (Scottish Natural Heritage (SNH)). From these findings and consultations, a scoping matrix was produced which simply considered direct/indirect impacts associated with the construction and operation of the bridge. At this stage in the EA, it was considered that the significant impacts were associated with changes in traffic flow that a third river crossing would produce, including positive impacts achieved by improvements elsewhere in the road network.

CASE STUDY: ALDER ROAD WEIR

The NRA (now part of the Environment Agency in England and Wales) was set up under the 1989 Water Act with the task of protecting and improving the water environment in England and Wales. Its duties include the regulation of discharges, protection from flooding, conservation and recreation. In carrying out some of these duties, the NRA has to promote certain works which have environmental effects (e.g. flood defences) and object to projects proposed by others where they are believed to compromise the NRA's duties regarding conservation and water quality. The NRA also has powers to authorise those features within a development which affect the water environment: for example, the granting of water abstraction licences. Therefore, within the EA process in the UK, the NRA may be acting as developer, objector, statutory consultee or planning authority depending on the nature of the project.

Table 2.3 gives an example of a descriptive matrix taken from guidelines on EA of land drainage schemes produced by Thames Region NRA (1989). This is a descriptive matrix, which considers impacts during the construction phase and end-state phase and a range of impact characteristics, namely: adverse/beneficial; strategic/local; long-term/short-term; intermittent/continuous; direct/indirect and irreversible/reversible.

TABLE 2.3A Descriptive matrix used for land drainage schemes
Construction phase: Alder Road Weir

Proposal	A/B	S/L	Lt/St	Int/ Cont	D/Id	I/R	Comments
Agriculture	—	—	—	—	—	—	
Amenity	A	L	St	Int	D	R	Construction activity will be disruptive.
Angling	—	—	—	—	—	—	
Aquatic biology	0	—	—	—	—	—	Maintain water levels during construction.

TABLE 2.3A continued

	A/B	L	Lt	Cont	D	R	
Archaeology	A/B	L	Lt	Cont	D	R	Existing structure of possible industrial archaeological interest. Archaeological watching brief.
Fisheries	0	—	—	—	—	—	Maintain water levels during construction.
Landscape	0	—	—	—	—	—	Avoid adjacent trees.
Planning	0	—	—	—	—	—	Local Planning Authority will need to be satisfied that residents have been consulted.
Recreation	0	—	—	—	—	—	Avoid blocking adjacent public footpath.
Water quality	—	—	—	—	—	—	
Wildlife	—	—	—	—	—	—	

TABLE 2.3B Descriptive matrix used for land drainage schemes
End state phase: Alder Road Weir

Proposal	A/B	S/L	Lt/St	Int/ Cont	D/Id	I/R	Comments
Agriculture	—	—	—	—	—	—	
Amenity	0	—	—	—	—	—	Assuming existing bridge is replaced to provide access to far bank for local residents.
Angling	—	—	—	—	—	—	
Aquatic biology	—	—	—	—	—	—	
Archaeology							
Fisheries	—	—	—	—	—	—	
Landscape	0	—	—	—	—	—	Sympathetic detailing on new structure in brick. (see Amenity).
Planning	0	—	—	—	—	—	
Recreation	B	L	Lt	Cont	Id	I/R	Assuming adjacent bridge of public footpath is refurbished.
Water quality	—	—	—	—	—	—	
Wildlife	—	—	—	—	—	—	

Key to headings:
A/B – Adverse/Beneficial
Int/Cont – Intermittent/Continuous
S/L – Strategic/Local
D/Id – Direct/Indirect

Lt/St – Long term/Short term
I/R – Irreversible/Reversible
(NB: 0: Neutral impact assuming comments noted are incorporated in final design)

Source: Thames Region NRA (1989)

Questions for thought

1. How important are methods in conducting EAs?
2. Describe how you would design a checklist, matrix and network diagram for one or more of the following development projects:
   - proposed landfill site;
   - shopping centre in a rural area on the edge of a town;
   - opencast mining development in an equatorial tropical forest;
   - power station on a coastal plain.
3. How would environmental features mapping methods be applied to a proposed road between the towns?
4. What are the advantages of using EA methods in site (alternative) selection studies?

Techniques for impact prediction and evaluation

◆ Introduction 34
◆ Air quality assessment 34
◆ Noise assessment 39
◆ Landscape and visual impact assessment 45
◆ Ecological assessment 52
◆ Water assessment 58
◆ Archaeological and cultural heritage assessment 64
◆ Social impact assessment 66
◆ Questions for thought 71

Introduction

EA techniques are employed to fulfil a number of functions although they are primarily used to predict and quantify the magnitude of impacts, evaluate and assess the importance of the identified changes, present information and monitor actual changes (Institute of Chemical Engineering 1994).

It is not the purpose of this book to appraise the full range of techniques used in EA. Nevertheless, it is important that the practitioner is familiar with the general principles and terminology of some of the commonly used techniques so as to have an appreciation of their purpose in the project cycle, their data requirements and the benefits of their use in the production of the ES. Commonly used EA techniques are:

- air quality assessment;
- noise assessment;
- landscape and visual impact assessment;
- ecological assessment;
- water assessment;
- archaeological and cultural heritage assessment;
- social impact assessment.

Technical assessments have been detailed for specific projects (e.g. roads and traffic (Department of Transport UK 1994)) and such guides have proved useful to ensure a degree of continuity and control over the technical assessment process. However, it is inevitable that the technical assessment process will differ from one ES to another. The disparate nature of differing scientific disciplines makes it difficult to standardise the process. Nevertheless, each of the techniques discussed below has a common theme. Firstly the existing baseline environmental setting is established, secondly the magnitude and significance of the impact is gauged and finally mitigation measures are recommended where appropriate to minimise impacts.

Air quality assessment (AQA)

Air quality impact

Air pollution can be defined as the presence in the external atmosphere of one or more contaminants (pollutants), or combinations thereof, in such quantities and of such duration as may be or may tend to be injurious to human health, plant or animal life, or property (materials), or which unreasonably interferes with the comfortable enjoyment of life, or property, or the conduct of business (Harrop and Carpenter 1992 and Canter 1996).

Assessment methodology

AQA is an air quality management tool which aids the efficient use of air resources. It involves not only identification, prediction and evaluation of critical variables such as source emissions and meteorological conditions, but also potential changes of air quality as a result of emissions from proposed projects and ultimately an assessment can be undertaken to ensure compliance with various ambient air quality standards (AAQS). The principal aim of AQA is to identify and quantify impacts and, through project design and planning, mitigate them to ensure that a development's impact is acceptable. Canter (1996) has identified six procedural steps for an AQIA: identification of air quality impacts of the proposed project; description of existing air environmental conditions; procurement of relevant air quality standards and/or guidelines; impact prediction; assessment of significant impacts; and identification and incorporation of mitigation measures. An AQA may, however, be broadly and simply divided into three stages (Harrop 1994, 1998) (see Figure 3.1).

The existing situation

The assessment begins with a knowledge of the existing situation. This will depend upon the ambient air pollution concentrations; pollutant sources and their specific location; meteorology; local topography; physical conditions affecting pollutant dispersion; and sensitive receptors and their specific location. The aim is to know what air pollutants are present in the area under consideration and in what quantities, where the pollutants came from, how they will be dispersed and where they are destined to impact upon a sensitive receiver.

Characterisation of emission sources

The second step is to determine the character of the released pollutant emissions (e.g. the nature of the pollutant; emission rate; efflux velocity; efflux temperature; and source morphology; etc.).

Assessment of impacts

The third stage is to review the impacts resulting from the identified emission source and where necessary mitigate the impacts. The assessment is generally based on a comparison of the AAQS (Murley 1995, World Health Organisation 1987a, 1995a and 1995b, Department of the Environment UK 1997) for the pollutant of concern and the

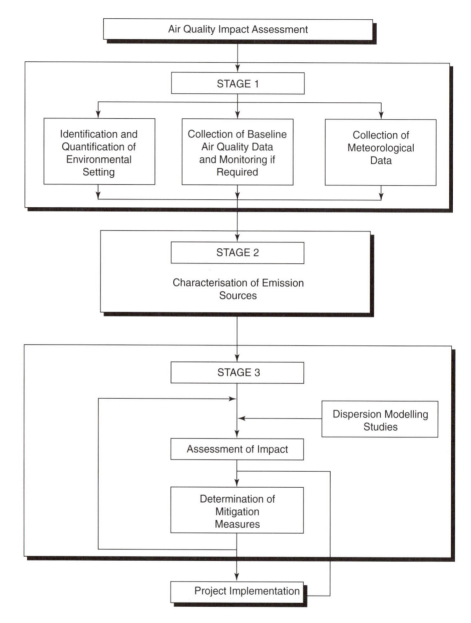

FIGURE 3.1 AQA procedures
Source: Harrop (1994), reprinted with permission of The Royal Society of Chemistry

cumulative concentrations (i.e. background and predicted incremental concentrations) of that pollutant. In order to avoid exceeding the AAQS, mitigation measures should be incorporated into the project at the design stage.

Computer-based models (e.g. ISCLT, COMPLEX, ADMS-2.2, PAL, AERMOD, etc.) are used to simulate the dispersion of air pollution into the atmosphere. The objective is to relate mathematically the effect of source emissions on ambient air quality and to establish whether permissible levels are, or are not, being exceeded. Models have been developed to meet these objectives for a variety of pollutants, time scales and operational scenarios. Short-term models are used to calculate concentrations of pollutants over a few minutes, hours or days and can be employed to predict worst-case conditions (e.g. high pollution episodes). Long-term models are designed to predict seasonal or annual average concentrations, which may prove more useful in studying health effects and impacts on vegetation, materials and structures. Szepesi (1989) provides a comprehensive description of more than 180 air dispersion models.

One type of model widely used is the Gaussian, where the spread of a plume in the vertical and horizontal direction is assumed to occur by simple diffusion perpendicular to the direction of the mean wind (Turner 1979 and 1994 and Pasquill and Smith 1982). The concentration of gas or aerosol at X, Y, Z from a continuous source with an effective emissions height, H, is given by:

$$\chi = ((Q)/(2\Pi\sigma_y \sigma_z U)) \exp\left(-\tfrac{1}{2}(Y/\sigma_y)^2\right)\{\exp[-\tfrac{1}{2}((Z-H)/\sigma_z)^2] + \exp[-\tfrac{1}{2}((Z+H)/\sigma_z)^2]\} \tag{1}$$

The notation used to depict the concentrations is the height of the plume centre line when it becomes essentially level and it is the sum of the physical stack height and the plume rise. The meteorological factors influencing plume rise are wind speed, u; temperature of the air, shear of the wind speed with height and atmospheric stability. Moses et al. (1964), having compared actual and calculated plume heights by means of six plume rise equations, reported 'There is no one formula which is outstanding in all respects'. Many formulae are available to derive plume rise estimates (Davidson 1949, Holland 1953, Bosanquet et al. 1950, Bosanquet 1957 and Briggs 1969) which give generally satisfactory results in test situations. Examples of air dispersion models commonly used in the UK include ADMS-2 (Carruthers 1995) and ISC3 (United States Environmental Protection Agency 1987).

To ensure the protection of human health, air quality criteria have been applied. The primary aim of air quality guidelines and standards is to provide a basis for protecting public health from adverse effects of air pollution and for eliminating, or reducing to a minimum, those contaminants of air that are known or likely to be hazardous to health. Guidelines and standards represent the current best scientific judgement, but there is a need for their periodic revision, since much remains to be determined regarding the toxicity of air pollutants for humans. Standards and guidelines may also be applied to the protection of flora and fauna and materials.

Many countries have their own national AAQS, but these can be limited in the scope of pollutants assessed. Therefore there is often a need to apply other national or international AAQS or, in the absence of suitable adopted standards, to derive them. For example, the only air quality standards in the UK that have any legal status are the Air Quality Standards Regulations 1997. They include:

- Particulate matter less than 10 microns in diameter (PM_{10})
- Carbon monoxide (CO)
- 1,3 Butadiene
- Benzene
- Ozone (O_3)
- Sulphur dioxide (SO_2)
- Nitrogen dioxide (NO_2)

- Lead

- 50 $\mu g \ m^{-3}$ 24–hour running average
- 10 ppm running 8–hour average
- 1 ppb running annual average
- 5 ppb running annual average
- 50 ppb running 8–hour average
- 100 ppb 15–minute average period
- 150 ppb average over 1–hr and 21 ppb annual mean
- 0.5$\mu g \ m^{-3}$ annual average

CASE STUDY: AIR DISPERSION MODELLING SCOPING EXERCISE (AFTER TURNER 1970)

An incinerator is proposed to be built and planners wish to know the possible impact of the development on air quality. It is estimated that 72 g sec^{-1} of sulphur dioxide (SO_2) would be emitted from a 45 m stack with a diameter of 1.5 m. The effluent gases will be emitted at a temperature of 395K with an exit velocity of 13 m sec^{-1}. The planners would like to know what the predicted ground level concentrations would be 500 m downwind of the stack on an overcast winter's morning with a surface wind speed of 6 m sec^{-1}. The atmospheric pressure is thought to be 970 mb and the ambient air temperature is 293K.

Using Holland's equation (2) the estimated plume rise is 8.1 m. Therefore the effective plume rise is the plume rise plus the stack height, which is 53.1 m.

Holland's equation is given as:

$$H = ((v \ d)/u)(1.5 + 2.68 \times 10^{-3} \ p \ ((T_s - T_a)/ \ T_s)d) \qquad (2)$$

(where: H is plume rise (m); v is stack gas exit velocity (m sec^{-1}); d is stack diameter (m); u is wind speed (m sec^{-1}); p is atmospheric pressure (mb); T_s is stack gas temperature (K); T_a is air temperature (K); and 2.68×10^{-3} is a constant having units of $mb^{-1} \ m^{-1}$).

To determine the ground level concentration (χ) in the aforementioned meteorological conditions downwind of the stack requires an abridged version of equation (1):

$$\chi = ((Q)/(\Pi\sigma_y \ \sigma_z \ U)) \ exp[-\tfrac{1}{2}((H)/\sigma_z)^2] \qquad (3)$$

The dispersion coefficients σ_y and σ_z may be determined from the graphs in Figure 3.2 or by the following equations: $\sigma_y = ax^{0.903}$ (4) and $\sigma_z = bx^c$ (5). Where x is

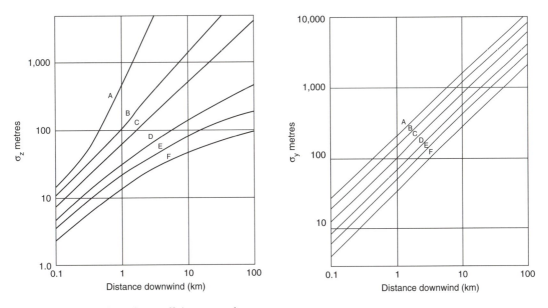

FIGURE 3.2 Dispersion coefficients σ_y and σ_z
Source: Turner (1994). Reprinted with permission by CRC Press

the downwind distance from the source, σ_y, σ_z and x are in metres. The parameters a, b and c can be gained using Tables 3.1 and 3.2 (Pasquill and Smith 1982). From Table 3.1 it can be seen that in overcast conditions with a windspeed of 6 m sec^{-1}, stability class D conditions (neutral) prevail.

From the information gained in Table 3.1 values a, b and c can be obtained from Table 3.2, where σ_y and σ_z become 35.6 and 17.9 m respectively. Therefore the predicted short-term (approximately 10 minute) ground level concentration is approximately 74 μg m^{-3}. The predicted level may then be added to the known short-term background concentration and compared with the recommended AAQS.

Noise assessment

Noise impacts

Apart from those people who are unfortunate enough to be profoundly deaf, we are all aware of sound in every moment of our lives. Sound allows us to enjoy music, and it enables us to communicate easily with each other. Noise is with us throughout our lives. It has been defined in many different ways according to the circumstances in which it occurs and the effect which it produces: 'a number of tonal components disagreeable to man and intolerable to him because of the discomfort, fatigue, agitation and, in some cases, pain it causes' (Commission of European Communities 1976) and 'sound which is undesired by the recipient' (Anon. 1963). All definitions indicate noise

TABLE 3.1 Pasquill stability categories

| wind speed | Daytime incoming solar radiation (m W/sq cm) | | | | Within 1 hour of sunrise or sunset | Night time cloud amount (oktas) | | |
	strong >59	moderate 30–59	slight <29	overcast		0–3	4–7	8
<2	A	A–B	B	C	D	F	F	D
2–3	A–B	B	C	C	D	F	E	D
3–5	B	B–C	D	D	D	D	D	D
5–6	C	C–D	D	D	D	D	D	D
>6	C	D	D	D	D	D	D	D

Source: Pasquill and Smith (1982)

TABLE 3.2 Fitted constants for the Pasquill diffusion parameters

| Class | $x \leq x_1$ | | | | $x > x_1$ and $\leq x_2$ | | | $x > x_2$ | |
	a	b	c	X_1	b	c	X_2	b	c
A	0.40	0.125	1.03	250	0.00883	1.51	500	0.00023	2.1
B	0.295	0.119	0.986	1000	0.0579	1.09	10000	0.0579	1.09
C	0.20	0.111	0.911	1000	0.111	0.911	10000	0.111	0.911
D	0.13	0.105	0.827	1000	0.392	0.636	10000	0.948	0.540
E	0.098	0.100	0.778	1000	0.373	0.587	10000	2.85	0.366

Source: Pasquill and Smith (1982)

as 'unwanted sound', either generally or by a particular person and/or in a particular time and place. It can, and does, cause hearing damage, but in the environment generally it seldom does any real physical harm. However, it has become one of the environmental issues that people most complain about. This is mainly due to an increased awareness of both problems and solutions as people begin to appreciate the meaning of health and safety in all situations, and the environment in general. Research clearly indicates its effects on hearing damage, disturbance of sleep, stress, annoyance and interference with communication. Because of its interaction with other environmental factors, however, it is sometimes difficult to analyse properly.

Assessment methodology

While the process of noise assessment appears to be straightforward, Canter (1990) has identified four issues that may tend to hamper the systematic addressing of noise in EA: confusing and overlapping terminology; logarithmic expression of noise levels;

lack of uniform environmental noise criteria or standards; and the paucity of noise data. While these issues are or may be a deterrent, they should not hinder the noise assessment process. The procedures for noise assessment are given in Canter (1990):

◆ identify noise levels for the alternatives under consideration during project construction and operational;
◆ determine existing baseline noise levels for the project area and identify unique noise sources in the area as well as noise sensitive receptors;
◆ obtain applicable noise standards and criteria for the area;
◆ determine the microscale impact by predicting anticipated noise levels for each alternative during project construction and operational. Compare predicted noise levels with applicable standards and criteria in order to assess impact; and
◆ if standards or criteria are exceeded, consider noise abatement methods to minimise impact on the noise environment.

Like many environmental disciplines noise assessment has a terminology that for many is difficult to comprehend. Principal terminology includes the definition of sound, which is a noise that can be heard. Strictly speaking, noise is any pressure variation that the human ear can detect. Sound is created by a vibrating object and the vibrations are transmitted by wave motion through air, liquid or solids to the ear. It cannot be transmitted in a vacuum. The speed of sound varies according to the medium in which it travels. If these variations in pressure occur rapidly, i.e. more than 20 times per second, they can be heard. The number of variations in a second is known as the frequency and is expressed in Hertz (Hz). At low frequencies air particles vibrate slowly, producing low tones, which are often difficult to control. If a sound occurs at only one frequency, it is called a pure tone. Pure tones are not common, but when they do occur in an environmental situation they can often be difficult to eliminate satisfactorily. To determine the composition of a sound it is necessary to determine the sound level at each frequency individually. The values are expressed in octave bands. The octave bands are referred to by their centre frequencies. For example, the 500 Hz octave band stretches from 354–707 Hz. Using sound level meters equipped with octave band analysers it is possible to finger-print the sound to determine the annoying frequencies and thus establish the appropriate method of control.

Sound is transmitted by a longitudinal wave motion, i.e. vibrations of air backwards and forwards parallel to the direction of wave travel. The wavelength is the distance from one wave to the next. Since the behaviour of sound is usually dependent upon its wavelength the relationship is important. The maximum movement of the medium up or down from its midway point is known as its amplitude and, generally speaking, the greater the movement the louder the sound. A sound at a constant pressure can be made to appear quieter or louder by changing the frequency.

A decibel is the unit of intensity of sound used to measure sound energy reaching the ear. The weakest sound that a normal human ear can detect is 20 millionths of a Pascal. This pressure change is so small that it causes the very delicate membrane of the human ear to be displaced by only a tiny fraction. The ear can also, however, tolerate sound pressure more than a million times greater. Measurement in Pascals would, therefore, produce unmanageable numbers so a simpler scale has been devised based on the decibel. It sets its reference point at the threshold of hearing (i.e. 20 Pa) and compresses the range of human hearing into a much more acceptable and manageable 120 dB range. However, the decibel scale is logarithmic and is, therefore, a scale not a unit. The human ear cannot respond with anything like the same sensitivity to all frequencies. When measuring sound it is important, therefore, to take this into account, so that the readings will reflect more readily the way the sound is heard. There are four internationally agreed weightings – A, B, C and D. It is an 'A' weighting network that conforms most to the way the human ear responds. Virtually all sound level meters are built with a filter to allow 'A' weighted measurements to be made. The 'A' weighting is intended as a method of making broad band loudness measurements. It takes much less account of the low frequency components of the noise than the mid-frequency and hence the fact that it corresponds more closely to the frequency response of the human ear. Noise measurements as A–weighted sound pressure are quoted in dB(A). The sound power level (SPL) is a measure of the total energy of the source. It is independent of environmental influences. The sound pressure level can be calculated at any distance from the source if the SPL is known.

Noise prediction and evaluation

Because it is difficult, if not impossible, to predict precisely the reaction of a particular individual to a noise there can be no guarantee that one set of control measures will improve a particular situation sufficiently to produce a satisfactory environment for all people exposed to it. When assessing a particular noise problem, therefore, it is essential that the following factors are considered: the physical characteristics of the source of the noise (level, tone, impulse, duration, intermittency and variability frequency spectrum) and the psychological circumstances of the receiver to the noise (personality, activity and situation). It is also important to remember, therefore, that the criterion of judgement is very much a matter of circumstance and that no simple single-figure measurement can be laid down to specify nuisance. It is also true that, to a remarkable extent, a noise, originally annoying or disturbing, can become tolerated and even unnoticed by most people when it becomes sufficiently familiar. Attempts to define criteria linking noise exposure and annoyance have led to the development of many methods for the measurement of both variables. This is particularly necessary as the impact of a noise is often dependent on its source. For example, it is known that a person will react to noise from a disco and to noise from road traffic in different ways. The reason for this lies with the components of the noise and not just the actual level. It is necessary, therefore, to ensure that when a noise is measured, this is done in a way that is particularly relevant to that noise.

Potential noise impacts arising from fixed items of plant can be assessed using the methodology recommended by BS4142: 1990. This methodology involves the comparison of predicted operational noise levels with existing noise levels at the facade of identified sensitive receptors. Assessments carried out in this way make it possible to predict the likelihood of complaints arising from operational noise. BS4142: 1990 states that noise from a proposed development which is 5 dB(A) higher than the existing background noise level is of marginal significance. A level which is, say, 3 dB(A) above the background would therefore offer a reasonable level of protection to local residential properties. This approach has been used elsewhere to allow a reasonable planning standard to be achieved.

The standard method for predicting noise impacts associated with open construction sites is BS5228: Part 1: 1984: 'Noise control construction and open sites'. Part 1 of this British Standard, 'Code of practice for basic information and procedures for noise control', provides a method for combining the contribution of noise from a number of individual items of plant, taking into account their locations, their sound power levels, and the percentage of the time that they are operating, or percentage 'on-time'.

Noise control and prevention of nuisance

Control of noise and the prevention of nuisance is often linked with ensuring that effective sound insulation and absorption techniques have been employed. This is especially important to remember at the design stage of any project. There is a clear distinction between insulation and absorption in respect to sound. A sound insulating material attenuates sound waves passing through and thus acts as a sound barrier. A sound absorbent material, on the other hand, absorbs a proportion of the sound emerging incident on it so that the level of sound reflected from the surface is substantially reduced. Sound absorption, therefore, reduces the loudness of reflected sounds in a room and decreases reverberation. It is essential, therefore, to identify the actual sound problem and the main acoustic weakness in each case before recommendations on sound control are made.

Noise control is basically a system problem. In general there may be, and often are, many components which can be manipulated to achieve a particular result. The system contains three parts: source, path and receiver. Control is normally directed to break this chain in some way to obtain an acceptable noise environment consistent with economic and operational considerations. The use of the word acceptable raises the questions of acceptable under what conditions? or acceptable to whom? There is, of course, no easy answer to these sorts of questions as each problem needs to be assessed separately, taking account of the complexity of considerations involved in each case. However, noise control is not necessarily synonymous with noise reduction. Although it is true that the overwhelming majority of noise problems are best resolved by effecting a reduction in the SPL, in some situations an appropriate solution might be to add another noise.

Every noise problem has its own peculiarities and to generalise would be, as already stated, very dangerous. However, in each case it is appropriate to consider the following options. Each option has the potential to mitigate noise and should be considered during an EA. These include: stop making the noise, remove the noise source, reduce the noise at source, screen the noise source or increase the distance from the source. Controlling noise is never straightforward and often the noise complainer is only satisfied by the complete elimination of the noise – an outcome that is rarely achieved. In extreme cases the removal of the complainer might be contemplated, especially where alternative housing accommodation is available.

CASE STUDY: NOISE ASSESSMENT OF AN INDUSTRIAL PLANT (SEE CHAPTER 9 (ASPINWALL & COMPANY 1995))

Noise at a proposed chipboard manufacturing plant was a potential cause of adverse impacts on the local community and environs. Noise was examined in relation to the possible impacts at locations identified as being potentially sensitive. Road traffic associated with the construction and operation of the proposed development, construction activities and the operation of the plant were assessed.

There were a number of existing noise sources in the vicinity of the proposed development. The significant ones were: a trunk road; the local road network; a paper mill; an industrial estate; a railway line; and the sea (see Figure 9.1 in Chapter 9). A baseline noise study was undertaken using monitoring locations and protocols agreed with the local Environmental Health Department. The background noise levels were:

◆ Night-time (Noise Sensitive Receptors (NSR1)): 40.0 dB $L_{A90(1\text{-hour})}$
◆ Night-time (NSR2): 40.0 dB $L_{A90(1\text{-hour})}$
◆ Daytime (NSR3): 40.0 dB $L_{A90(1\text{-hour})}$

The number of additional vehicle movements associated with the construction and operation of the site was sufficiently small that they would not make a significant contribution to noise at NSRs. The proposed plant was laid out such that most items of fixed plant were situated towards the northern end of the site away from the NSR. Laying the plant out in this way resulted in a greater distance between the nearest residential properties and the noisiest items of fixed plant, and ensured that these properties also benefited from a degree of acoustic screening in respect of fixed plant. The operation of the fixed plant would occur uniformly 24 hours per day, including weekends. The night-time situation therefore represented the worst case at the NSR as existing background noise levels were lowest during the night. It was also necessary to consider the noise impacts generated due to mobile plant and vehicle movements on the site during the daytime. These sources included delivering vehicles; trains entering and leaving the site via a proposed rail link; mobile plant involved in the handling of materials in and around buildings; and employees' vehicles using the car park. The predicted worst-case noise levels at the NSR during the day and at night from fixed and mobile sources were:

◆ Daytime:
    NSR1 due to all on site noise sources 50.5 dB $L_{aeq(t)}$
    NSR2 due to all on site noise sources 34.5 dB $L_{aeq(t)}$
    NSR3 due to all on site noise sources 47.1 dB $L_{aeq(t)}$

◆   Night time:
       NSR1 due to all on site noise sources 39.7 dB $L_{aeq(t)}$
       NSR2 due to all on site noise sources 33.2 dB $L_{aeq(t)}$

Predictions showed that noise from the operation of the plant would be of less than marginal significance with respect to generating complaints. Operational noise levels due to the combined noise sources on site were below the limits proposed.

Landscape and visual impact assessment

Landscape includes aesthetic, cultural and amenity as well as physical components – a fact which inevitably leads to the inclusion of both subjective and objective means of assessment. Visual impact assessment tends to be more objective than landscape assessment since techniques are largely concerned with the extent to which a proposed development is visible rather than with any quantification of human reaction to, or perception of, the visual intrusion of a development. Landscape assessment can be applied at the project, plan and policy levels, whereas visual impact assessment is project specific.

The Countryside Commission approach to landscape assessment (Countryside Commission 1987) was originally developed for its work in relation to designated landscape areas but is of use in a wider context. The approach has objective and subjective elements in view of the fact that the Commission is concerned with the appearance of the land and people's reactions to it. The following approach is adapted from the Countryside Commission (Countryside Commission 1991) and applied to the requirements of an EA.

Scoping

Scoping of landscape impacts should take into consideration particular features of the proposed facility, its size and duration of operations together with aspects of the local environmental setting. These include local landscape resources, the quality of views, components of the landscape which are protected (including cultural and historical features), and proposals for development in local plans which could bring additional elements into the landscape (Petts and Eduljee 1994). As well as any statutory consultees, national bodies, local amenity and conservation groups and representatives of the local community should be consulted to establish (and take into account) issues of concern.

Description of the development

Specific aspects of the development should also be described where these have an impact on the local visual and landscape character (positive and negative impacts). It may be appropriate to consider these in terms of the phases of development (construction, operations, post-operations, restoration). The Countryside Commission Guidelines identify two elements of a project which are likely to be relevant to potential landscape impact: the appearance and layout of the main elements of the facility, including size, materials, colours and forms; and non-visual characteristics of the project, including emissions of all types (e.g. noise) which would be of importance to human perceptions of the landscape. The latter have received little attention in ESs prepared in the UK, a situation which illustrates a limited understanding of the concept of landscape and the tendency to ignore interactions between environmental factors (Petts and Eduljee 1994).

Baseline assessment

A description of the existing landscape character and visual amenity within the vicinity of the proposed development is made which notes features of particular landscape interest such as rivers, mountain ranges, woodlands and designated areas of landscape importance. The assessment may be either a desk study or a field study. The former involves a review of the existing literature such as previous landscape assessments, reference to site designations and guidebooks. Maps can be used to determine the most appropriate routes for conducting field surveys. Aerial photographs, when available, can also be used with some effect. Table 3.3 provides an example of the type of field survey form used by the Countryside Commission for landscape assessment. Sketch maps, photographs and map annotations can add to the value of the field data obtained. It may also be useful to include information on the condition of the landscape, for example, the age of woodlands, species composition, etc.

Impact prediction and significance

Examples of individual landscape impacts (positive, negative, temporary, permanent, etc.) whose magnitude may need to be predicted are as follows:

◆   the area of landscape directly occupied by the project and how it is affected;
◆   the zone of visual influence of structures/ land use changes associated with the project;
◆   the number and ways in which visual or functional landscape features are affected by the project;
◆   the overall effect of the project on the perceived landscape character;
◆   the impact on other environmental components which are important determining factors in the overall character of the existing landscape;
◆   the impact on the direction and rate of change in the landscape which would be expected to occur in the absence of the project.

TABLE 3.3 Field survey sheet for landscape assessment

Project: Surveyor:
Date: Time: Weather:
Viewpoint: Direction of view:
Description: general impression
Any significant seasonal differences
Sketch

1. Objective checklist
Record what is present by marking relevant words:
– inconspicuous   X evident   XX conspicuous
Landform
flat   plain   coast
rolling   rolling lowland estuary
undulating   plateau   broad valley
steep hills   narrow valley   vertical crags
deep gorge
Land cover
built-up   arable   deciduous wood   marsh   cliff
road   pasture   coniferous wood   river   beach
industry   moor   mixed wood   lake   dune
mineral working   scrub   parkland   reservoir   mudflat
Landscape elements
farm buildings   walls   woodland   river   footpath
churches   fences   plantation   waterfall   track
ruins   hedges   shelterbelt   rapids   road
masts, poles   banks   tree clumps   falls   motorway
pylons   isolated trees   pond   railway
                                     hedgerow trees   canal   car park

2. Subjective checklist
Record your immediate impressions by marking each line with a circle around or nearest to the most appropriate word.

Scale	intimate	small	large	vast
Enclosure	tight	enclosed	open	exposed
Variety	uniform	simple	varied	complex
Harmony	harmonious	balanced	discordant	chaotic
Movement	dead	calm	busy	frantic
Texture	smooth	managed	rough	wild
Colour	monochrome	muted	colourful	garish
Rarity	ordinary	unusual	rare	unique
Security	comfortable	sage	unsettling	threatening
Stimulus	boring	bland	interesting	invigorating
Pleasure	offensive	unpleasant	pleasant	beautiful

Source: Countryside Commission (1987)

In general, the significance of a landscape impact(s) will be determined by the quantity and quality of landscape affected. This means that an impact in an area noted or designated for its landscape value is going to be more significant than a landscape change occurring elsewhere. However, it should not be forgotten that a change in a local, undesignated site may be of major local significance. Where landscapes affected by a development have not been granted any official landscape designation, criteria such as some of those used in ecological assessment may be used in order to assess the relative significance of an impact on the landscape. These include rarity, representativeness and sensitivity to change.

Evaluation of landscape quality

In the UK, the effort and research in landscape evaluation has concentrated on the production of objective and quantifiable methods, using statistics which attempt to reduce the complexities of, and emotive responses to landscape assessment (Countryside Commission 1988). A number of methods have been developed based on field and measurement methods (Martin 1985). Evaluation methods involve professional judgement and/or public preference.

In visual impact assessment, emphasis is placed on defining visual envelopes (zones of visual influence (ZVI)) and presenting graphic or other illustrations of how the visual impact is perceived by those living in the vicinity. The ZVI is the area surrounding the proposed development from which the site is visible. Two other terms need to be distinguished: visual intrusion and visual obstruction. Visual intrusion is the extent to which the proposed development intrudes in the surrounding landscape and is dependent, to an extent, on the quality and type of landscape within which the development is located (Department of Transport UK 1983 and 1994). Visual obstruction occurs when the view is appreciably cut off from an observer. This has previously been expressed as a three-point scale (slight, moderate or high) (Department of Transport UK 1983).

In determining the visual impact of a development, account should be taken of the visual influence of land forms surrounding the site and other variables, for example, changes in seasonal, weather and lighting conditions (Martin 1984). A range of techniques are available to illustrate graphically the location and visual/landscape impact of a proposed development. These include low-tech plans, perspective sketches and physical models and the more high-tech approaches employing digitised data. These techniques are essentially objective and concerned with defining the extent to which developments are visible without attempting to quantify human reaction to or perception of the intrusiveness or otherwise of the development. A selection of techniques include: photography (Sampson 1992 and Smarden et al. 1986), terrain modelling and computer painting. The choice of techniques and level of detail, which can be obtained through landscape assessment and visual impact assessment, are extensive. At present the emphasis is towards visual representation with little account for public perception. In the future, greater emphasis will need to be placed on the inclusion of public preferences in the decision-making process.

Guidance on how landscape and visual impact assessments should be carried out has recently been produced (Institute of Environmental Assessment 1995).

CASE STUDY: ASSESSMENT OF A PROPOSED CHIPBOARD MANUFACTURING PLANT

A recent assessment of the magnitude and importance of positive and negative landscape and visual impacts associated with a proposed chipboard manufacturing plant (Aspinwall & Company 1996a) used the following terminology to describe the impact of the plant (see also Figure 3.3 and Table 3.4):

TABLE 3.4 Landscape and visual sensitive receivers

Sensitive receivers	Magnitude of impact	
Residential	*Construction stage*	*Operation stage*
1. Treeshill	Severe	Severe
2. Knagshill	Substantial	Substantial
3. Meikle Heateth	Moderate	Moderate
4. Little Heateth	Substantial	Substantial
5. Knowe	Slight	Moderate
6. Broom Crescent, Ochiltree	Minimal	Slight
7. Stewart Avenue, Ochiltree	Minimal	Slight
8. Douglas Brown Avenue, Ochiltree	Minimal	Slight
9. Netherton, Ochiltree	Minimal	Slight
10. Arran Drive, Auchinleck	Minimal	Slight
Roads and railway		
11. A76(T): East of Brackenhill A–B	Minimal	Slight
12. A76(T): Crossing B7036	Slight	Moderate
13. A76(T): North of Roseburn Roundabout	Slight	Slight
14. A70: Ochiltree–Humeston Wood	Minimal	Slight
15. A70: South-west of Ochiltree	Minimal	Slight
16. B7036	Minimal	Slight
17. B7046	Substantial	Moderate–Slight
18. Minor Road Scarea	Minimal	Minimal
19. Minor Road Burnock Bridge A70	Minimal	Slight
20. Kilmarnock–Gretna Green Junction railway line	Minimal–Slight	Slight–Moderate
Institutional		
21. Auchinleck Academy	Minimal	Minimal

Note: Comments relate to Figure 3.3

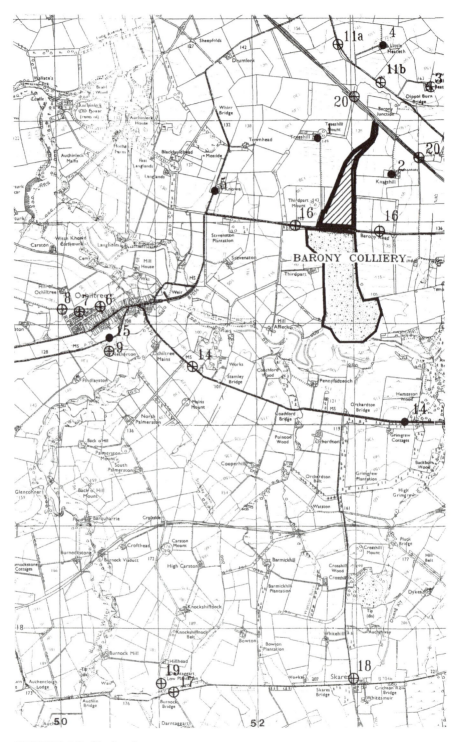

FIGURE 3.3 Site location map

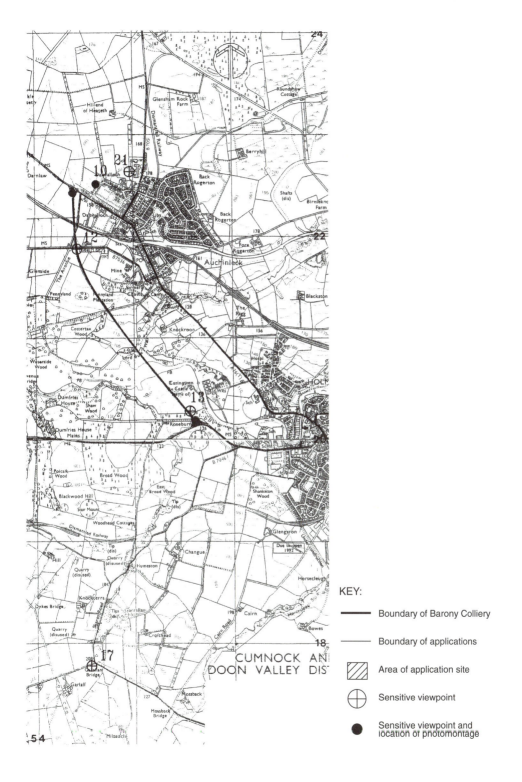

KEY:

——— Boundary of Barony Colliery

——— Boundary of applications

▨ Area of application site

⊕ Sensitive viewpoint

● Sensitive viewpoint and
location of photomontage

Minimal effect: where the site occupies only a small fraction of the overall view and is at a considerable distance from the viewpoint, and where the change to the landscape setting is minimal;

Slight effect: where only a minor portion of the view changes for a short duration or where the change to the landscape setting is minor;

Moderate effect: where some changes occur in the view or landscape setting, but not for a substantial length of time or not in a substantial part of the view, or where the change to the landscape setting is not substantial;

Substantial effect: where changes in the view or landscape setting substantially alter the overall scene or cause some alteration of the overall scene for a substantial length of time; and

Severe effect: where the overall view and landscape setting are altered for much of the operation period.

Ecological assessment

Ecological impacts

Ecological assessment is concerned with the study of the effects of development projects on the natural environment in relation to changes in habitat, the composition/population density of plant and animal species and ecological processes (such as nutrient cycling, energy flow, primary and secondary production, eutrophication, and succession). While negative impacts involving habitat damage/destruction and species loss/displacement often dominate the findings of an ecological assessment, other more positive impacts may also arise, for example the creation of new habitat during the restoration of a mineral extraction site, leading to greater biological diversity.

Assessment methodology

A wide range of ecological impacts are associated with development proposals and these occur over varied spatial and temporal boundaries. The complexity of ecological systems often makes the measurement and prediction of impacts (particularly indirect impacts) a difficult task. The assessment normally begins with a review of existing information sources such as maps, aerial photographs, local authority plans and previous ecological surveys or monitoring data. This enables the location and character of features of ecological importance such as woodlands, hedgerows or wetlands to be determined. Some countries have established ecological data centres but there are many more where ecological data are not so readily available. Where limited or no published data bases exist, great importance should be placed on seeking out members of the local community who have a good knowledge of their natural surroundings.

In the case of the British Isles, a wealth of information on the fauna and flora has been generated by countless forays by amateur naturalists and research students and by surveys carried out by national/regional conservation organisations such as English Nature, SNH, the Institute of Terrestrial Ecology (ITE) and the wildlife trusts. The Biological Records Centre (BRC) holds ecological data bases, remotely sensed data on land use and vegetation and GIS. On a local level, the wildlife trusts and Local Biological Records Centre have an important role in analysing and providing ecological data. The operations of these organisations have been described by Appleby (1991) and Harding (1991).

It is normally impractical to obtain data for all taxonomic groups since this is too time consuming and expensive. Therefore decisions have to be taken on the selection of the most important or appropriate species or groups of species. This may be based on the international/national/local importance of species and habitats in relation to their protected status (e.g. species included in Red Data Lists: Shirt 1987, Batten *et al.* 1990, Bratton 1991 and Perring and Farrell 1983). More pragmatically, it is those species which are more easily identified that are concentrated on in ecological assessments: flowering plants, birds (Baillie 1991) and butterflies (Pollard 1991) are the most popular. Alternatively, indicator species may be used to evaluate the environmental quality of a site (e.g. lichens for assessing air pollution, e.g. SO_2).

Biotic indices have been developed to measure the response of key species or groups to pollution and their use provides a cost-effective approach to ecological assessment, particularly in relation to impacts on aquatic systems. A number of such biotic indices have been developed in the UK, the most widely adopted being the BMWP score (Biological Monitoring Working Party) (Anon. 1981). Using a standard sampling method for aquatic macroinvertebrates (the 'kick sample') (International Standards Organisation 1985), families of macroinvertebrates present are scored and summated to produce a site score. Score values for individual families reflect their pollution tolerance (pollution intolerant families score the highest). An indication of organic pollution is given by dividing the score by the total number of scoring taxa to produce an average score per taxon (ASPT). A high ASPT value represents a relatively clean water community. Conversely, low scores denote more polluted areas.

The aim of an ecological baseline study is to describe the ecological conditions without a development proceeding in sufficient detail for subsequent predictions of conditions associated with the construction, operational and post-operational (where appropriate) phases of the development to be made. Baseline studies usually consist of a combination of desk studies, field survey and data analysis.

In Britain, a standardised approach for assessing terrestrial and freshwater systems is evolving based on Phase 1 habitat survey (Nature Conservancy Council 1990), followed by a more detailed Phase 2 habitat survey using the National Vegetation Classification (NVC) (Rodwell 1991a, 1991b, 1992, 1993, 1995) for targeted locations. The results of this initial survey together with the responses received from consultees may warrant the commissioning of more detailed studies of sensitive habitats or individual species. A complementary procedure for assessing river corridors has also been published (National Rivers Authority 1992).

An important benefit of introducing this standardised system for classifying and mapping wildlife habitats is that surveys may be carried out to a consistent level of

TABLE 3.5 Plant communities of the National Vegetation Classification

Aquatic (A)	Heath (H)
Swamp (S)	Woodland and scrub (W)
Mire (M)	Salt marsh (SM)
Mesotrophic grassland (MG)	Sand dune (SD)
Calcicolous grassland (CG)	Maritime cliff (MC)
Upland and acidic grassland (U)	Weed community (WD)

detail and accuracy and can be compared with one another. All land within the designated survey area is visited or inspected by a surveyor trained in using the technique. Vegetation is mapped, usually onto a 1:10000 Ordnance Survey (OS) map using around ninety specified habitat types. Each of these is assigned alphanumeric, lettered and colour codes which are recorded onto the map. It is not normally necessary physically to visit every part of the survey area. Field work can be augmented by the use of aerial photographs. Reference should also be made in the survey to topographic and substrate features. This is particularly important where vegetation is not the dominant component of the habitat. The production of target notes is an important part of the Phase 1 survey. The more comprehensive these are, the more useful the survey becomes. Target notes provide, for example, further information on sites of ecological importance such as rare species and site management and aim to provide at least a preliminary assessment of the nature conservation value of a site. A standardised procedure must be followed to achieve a reliable classification. This involves the quantitative description of all plant species for a suitable number and size of quadrats (samples). NVC keys published in each volume of British Plant Communities (Rodwell 1991a, 1991b, 1992, 1993, 1995) can be used to classify a plant community or sub-community (Table 3.5). Alternatively, the computer program MATCH (Malloch 1990) can be used. This compares the constancy values (frequency of occurrence) of field data with the constancy profiles of the NVC communities, from which a selection of community type is made.

River corridor surveys (RCS) were developed by the National Rivers Authority (now part of the Environment Agency for England and Wales) as a standard method of ecological survey in order to highlight important features in need of protection and to identify opportunities to rehabilitate and enhance degraded habitats (National Rivers Authority 1992). A river corridor consists of a stretch of river, its banks and the adjacent land (about 50 m wide). The RCS is based on recording major habitats, vegetation and physical features rather than detailing species and community records and is therefore more closely related to a Phase 1 habitat survey. The standard RCS deals with a 500 m stretch of river and maps onto a 1:2500 scale OS map. Information is recorded for four zones (Table 3.6) on to annotated maps using standard symbols and definitions. Accompanying site descriptions include features of the river channel, bank zone habitats, adjacent land use, details of species of special interest, recreational features, and management, such as bank mowing.

TABLE 3.6 River corridor survey zones

Zone	Detail to be recorded/mapped
Aquatic	plant communities
	flow and current features
	substrate and physical features
Marginal	plant communities
	substrate and physical features
Bank	tree species
	other plant communities
	physical features
Adjacent land	habitat types
	land use

Source: NRA (1992)

Decisions on the range of ecological impacts to be included in an assessment are usually made using comments received from conservation organisations/agencies as a guide. These might highlight notable rare species believed to occur within the development site and requiring detailed population surveys, or the presence of habitat at risk from damage by specific development actions such as drainage, topsoil disturbance, burning and so on. A useful approach is to consider the impacts occurring at each development phase as used elsewhere in EA (Walsh *et al.* 1991).

Ecological surveys inevitably generate large amounts of data and it is important to plan how these are to be managed and interpreted using appropriate statistical techniques. Increasingly, multivariate techniques involving ordination and classification are being used, particularly where species–environment interactions are being investigated. A useful introduction to this subject is given by Kent and Coker (1992).

Ecological impact prediction and evaluation

Predictions of ecological change as a consequence of development are generally more difficult to make than, say, changes in air quality or noise levels, mainly because of the inherent complexity and variability of ecosystems compared with physico-chemical systems. Perhaps as a result of this, ecological predictions are not always made quantitatively, but tend to be based on value judgements formulated from literature reviews (including ESs for similar projects and habitats), comparisons with species records for similar sites and developments, and expert opinions.

One relatively simple way of quantifying the magnitude of habitat damage or loss is to use an overlay technique (see Chapter 2). An overlay of the development plan is placed over habitat maps and the percentage habitat loss/damage caused by buildings, infrastructure, etc. is calculated. Models can be used to help predict ecological change. Probably the greatest efforts in quantitative modelling have been with faunal species. An example is the impact of estuarine tidal barrages on wader bird communities. Goss-Custard *et al.* (1991) suggested that post-barrage population densities of waders can be predicted from post-barrage population densities of certain prey species such as the polychaete worm *Nereis diversicolor*, which in turn can be predicted by reference to changes in sediment mixing and stability, water turbidity and other factors. Ideally, whenever ecological models are used in impact prediction, their accuracy and precision should be tested by means of operational/post-project monitoring programmes. This allows model refinements to be made where significant discrepancies between prediction and reality arise.

While a number of ecological and conservation evaluation methods exist, they are infrequently used in EA (Spellerberg 1992). The significance of a loss of species or habitat is usually evaluated in terms of its local/national/international conservation value, using (in the case of the UK) criteria such as species richness and diversity and habitat rarity, typicalness and fragility as described by Ratcliffe (1971 and 1977). These were originally developed for use in selecting nature reserves and have also been used in the formulation of conservation management plans. Other criteria for evaluating ecological significance include resilience and combined terms such as species security (measures of rarity and threat to specific impacts) (Treweek 1995).

Methods for quantifying some of these conservation criteria such as rarity have been devised for plants (Dony and Denholm 1985) and invertebrates. One example is the site rarity index calculated for ground beetles (Carabidae) in north-east England (Eyre and Rushton 1989). Comprehensive species lists for a range of sites (2 x 2 km tetrads) were compiled based on pitfall trapping methods (Eyre *et al.* 1986). From these data, species rarity values were assigned on a seven-point scale according to the number of sites in which they were found and a site score produced as the sum of scores for each species. This index (the beetle rarity total, BRT) was scored on a geometric scale (i.e. 64, 32, 16, 8, 4, 2, 1) instead of seven to one. From these scores (from all tetrads) a species rarity total (SRT) was calculated. A third index, which quantified rarity association, was also calculated to help identify sites containing several rare species. The rarity association value for a site was calculated using species that scored two or more in the geometric scale and corrected to eliminate bias caused by the presence of one very rare species in a list. For each of the three indices, the totals were then divided by the number of species in the list to produce three quality ratings: beetle quality factor (BQF), species quality factor (SQF) and rarity quality factor (RQF). There are a number of benefits to be gained from using an approach such as this. The site assessment is quantified rather than relying on a value judgement. The index provides a useful summary of potentially complex data sets for decision makers who are not necessarily experienced ecologists but need to know the relative ecological quality of a range of sites. The index is based on rarity, which is regarded as the most political of conservation criteria (Usher 1986) in that public understanding of it is greater than that of any other criterion. Finally, the index is based on a single group of invertebrates which is relatively easy to record, and can be applied to other groups. Such an assessment of site quality, however, should be complementary to, not instead of, a vegetation survey.

Once an ecological assessment has identified and/or predicted significant impacts associated with a development proposal, a strategy for preventing or mitigating these impacts should be devised. However, it is important to make mitigation relevant and specific to the ecology of species and habitats in question and the intended after-use of the site. Recommendations for mitigation would normally follow consultations with appropriate conservation bodies, both official and non-governmental organisations (NGOs).

Site restoration is a common component of the ecological mitigation proposals set out in ESs. The measures to be adopted should ideally be described in a restoration management plan. This would include details of planting proposals, faunal reintroductions (if any), soil and water management and an after-care programme including monitoring of the effectiveness of the restoration. Often, a site restoration is required which not only retains but enhances the conservation value of a site. Some general guidance on habitat restoration is given by Emery (1986), and for some specific development types by Coppin and Bradshaw (1982), the Department of the Environment UK (1986) and Andrews and Kinsman (1991). Ecological assessment should, by definition, deal with the impacts of development proposals on the natural environment, i.e. flora, fauna and on ecological processes. While there are examples of good practice, reviews of the quality of ecological assessments in Canada (Beanlands and Duinker 1983) and in the UK (Spellerberg 1994) indicate that there is some scope for improvement. Many ESs contain little more than lists of species found, possibly together with distribution maps, and do not include studies of ecological processes. Often, insufficient attention is paid to analysing data and interpreting their significance. There is also a need to strengthen predictions of ecological impacts and judgements on their significance by quantifying these whenever possible and testing through post-project monitoring (see Chapter 7). Recommendations for mitigation should be appropriate to the specific project and not couched in general terms.

A number of measures to help improve the quality of ecological assessments have recently been introduced in the UK. A standardised approach to vegetation description in baseline studies as described here is being widely adopted. The Institute of Environmental Assessment has produced guidelines on good practice in ecological assessment (Institute of Environmental Assessment 1995).

CASE STUDY: NORTH WESTERN ETHYLENE PIPELINE ECOLOGICAL ASSESSMENT

The North Western Ethylene Pipeline ES (Shell Chemicals UK Ltd 1989) was supported by a comprehensive series of reinstatement recommendations (Aberdeen Centre for Land Use 1989). This pipeline was designed to meet rising demands for ethylene as a feedstock for the petrochemical industry. The 406 km preliminary route was inspected by helicopter and filmed on video to ascertain environmental and engineering constraints before selecting a final preferred route. This film identified sites within a 2 km strip centred on the pipeline, which might be sensitive to disturbance and were considered important for their scientific or conservation value.

TABLE 3.7 Ecological assessment and reinstatement recommendations for a site forming part of the North Western Ethylene Pipeline

Site 1 (Avon riverbanks)
Location: KP 1.26 (0.03 km)

Vegetation summary:	Reinstatement recommendations:
(a) Tall herb grassland with scrub	3.1
(b) Scrub willows and weeds	3.20

Prescription

These small, high conservation value sites should be reinstated by careful turf stripping and replacement. Work should be prior to May or delayed until mid-July or later to permit most species to complete the phase of active growth. Reinstatement by seeding would be difficult to achieve, as grasses would be likely to become dominant, at least in the short term.

(1) Use a narrow easement
(2) Cut vegetation to 100m and remove debris, retaining about 20% for rush suppression (see below)
(3) Strip turf to a depth of at least 200 mm
(4) Store turf on geotextile sheets for not more than 3 weeks; keep moist
(5) Strip topsoil and store separately from subsoil as per normal practice
(6) Replace soil and turf in sequence. Close butt turf and pack gaps with topsoil
(7) Top-dress with 50 g/m^2 Enmag (4.5:20:10) low nitrogen slow-release fertiliser
(8) No seeding should be required unless substantial areas of bare ground remain, when the following mixture (which is intended mainly as a nurse crop) can be used:

Holcus lanatus	90%	} 4g/m^2 (40 kg/ha)
Agrostis capillaris 'saboval'	10%	

Source: Aberdeen Centre for Land Use (1989)

These sensitive sites were further examined in a second survey, the results of which led to a number of minor route deviations.

Information for reinstatement recommendations was obtained from field surveys which included: boundaries of major plant communities within each site and the species composition of each community; species records from permanent transects established in a representative selection of sites; and environmental data including details of the seed bank, soil type and chemistry, water table and rooting depth, site management (if any), slope, drainage and vegetation height. Reinstatement prescriptions were then written for each site, emphasising greater care for those with a high conservation value (Table 3.7).

Water assessment

Water impacts

Water impacts may relate to either surface and/or groundwater. Quantitative measurements of water quality and quantity are necessary to permit the assessment of impacts from a proposed development. Measurement of these pollutants is, however, fraught with difficulties. The measurement of water quality is problematical because specific materials responsible for the pollution are sometimes unknown; and pollutants are generally low in concentration and very accurate methods of detection are required. A guide to analytical techniques used in water and waste water engineering has been compiled by the American Public Health Association (American Public Health Association 1980).

Pollutants alter aquatic ecosystems in three main ways (Clark *et al.* 1981):

◆ through reducing concentration of dissolved oxygen. Oxygen content is the simplest single criterion for water purity, as most desirable species of aquatic life require oxygen to sustain life and activities;
◆ through directly causing death or reducing reproduction potential;
◆ by alteration of habitat or interference with food webs. For example an increase in settleable solids may alter the bed of the waterbody causing growth of plants. The resultant changed habitat may be unsuitable to certain desired species.

Surface water assessment

Water pollution can be defined in a number of ways, but the basic features of most definitions address excessive concentrations of a particular substance or substances for sufficient periods of time to cause a detrimental or identifiable effect. Water quality represents a term associated with the composite analysis of physical, chemical, and bacteriological parameters.

Prior to undertaking a water pollution impact study it is important to formulate a conceptual framework of the project (Canter 1983). In an assessment of the impact of contaminants on waterbodies there are two distinct stages: an accurate assessment is required of concentrations of induced pollutants and the time period during which they may be in contact with sensitive species or communities; and a consideration should be made of the effects of these concentrations on aquatic life and on human beings who consume or otherwise utilise the water. In some respects these effects can be predicted accurately as they are based on biological and toxicological knowledge. Questions of amenity, visual impact and land value are less easy to evaluate (Clark *et al.* 1981). Information will also be required on the development as well as the emitted pollutants, and a description of existing water quality conditions. In the initial stages of the study it may be necessary to conduct field studies and consult with organisations (e.g. the Environment Agency) and pressure groups with concerns for the environment to elicit views and possibly information.

A large amount of data is required to assess the impact of water pollutants. Required data should refer to aspects concerning human exploitation of, and relation with, the waterbody in question. When eliciting advice on possible effects of aqueous pollutants the planning authority and/or developer should (with others) evaluate both existing and predicted use because conflict may arise between polluters and future water users. Water intended for industry may have to meet required standards if it is to be used for certain processes such as the preparation of food. Plans for the future use of the waterbody should also be consulted, as planning approval may affect their implementation. Similarly, the proposed effluent should be related to possible environmental improvement programmes, such as riverside walks, which the planning authority may wish to implement. Close consultation between the planning authority and statutory consultees and other interested parties concerned with the waterbodies and its environmental setting is, therefore, very important.

Prior to an impact assessment of a new discharge, an identification of actual and likely sources entering or likely to enter the water course is an important part of the study of establishing the existing situation or baseline conditions. Water contaminants can enter a waterbody from a number of sources. Broadly these can be categorised into point and non-point sources. Impacts on waterbodies may arise not solely from discharges of waste effluents but also from activities concerned with the construction of the proposed development. The EA should also obtain information on the importance of the waterbody for fishing (commercial and/or recreational), agricultural supply, recreation (types of activities and number of participants), visual amenity, conservation policies and scientific research, and should identify all sites of conservation value dependent on the waterbody, officially and unofficially recognised, and assess the merits of each of these. Species as well as habitats must also be taken into account. Wildlife dependent on a waterbody must be considered if associated with an aquatic system likely to receive effluent. Local natural history societies and national organisations can often provide valuable advice and data.

Unfortunately the complexity of water pollution is exacerbated by the divergence of types of aquatic systems, namely: flowing freshwater surface systems, such as rivers; static freshwater surface systems, such as lochs; estuaries; coastal waters; and underground waters. The possible effects of water contaminants for all waterbody types can be: reduction of solar energy available in the ecosystem; increased input of nutrients, stimulating the growth of undesirable species, which may replace desired ones; reduction in availability of nutrients through increased sedimentation and neutralisation through adsorption causing inhibited growth rates; creation of intolerable physical extremes or ranges for some species, e.g. heat; elimination of reduction in success of species and individuals by toxicity; reduction in species diversity; interference with energy flow by materials that inhibit or alter feeding patterns; interference with decomposition and release of nutrients; decreased total weight of living material, biomass, by reduction of abundant species; and increased biomass by removal of important consumers, e.g. the loss of fish may lead to runaway production of other species due to reduction in predation.

Predictive models used to determine the concentrations of pollutants in receiving streams depend on many physical, chemical and biological parameters concerning the water and its discharge. The relative importance of the various natural processes varies widely both within and between similar bodies of water. Data requirements include flow regime, volume and location of discharge(s), coupled with the physical, chemical and biological processes and quality parameters of the receiving stream, thermal and chemical stratification and biological characteristics. Predictive methods may include usage of mass balance approaches or specific numerical models.

Qualitative projections can also be utilised. The objective is to delineate anticipated changes in baseline conditions that would result from the proposed developments. Having established a possible impact, it is important to establish the significance of the impact and identify mitigation measures to reduce the impact where possible.

Groundwater assessment

The development of water resources projects can also cause certain undesirable impacts on groundwater resources. In considering the potential impacts of a variety of project types on groundwater resource, attention should be given to both quality and quantity impacts. Similar consideration should also be given to surface water issues.

One major effect of development on catchments is to truncate the groundwater/ spring flow relationship. Under normal conditions incidental rainfall is intercepted by vegetation cover; part flows as surface run-off to streams and other major drainage channels while part slowly percolates through the soil and charges the groundwater. Lateral flow or seepage of this groundwater then helps to maintain stream and river flow. One effect of development is to disrupt this balance. Initially, vegetation cover is removed and replaced by a variety of hard surfaces. This results in little infiltration or percolation, most of the incidental rainfall being discharged as surface run-off. If this is discharged through stormwater drains to outfalls in streams, stream flow may exhibit typical surges after rainfall. The dampening of oscillations in flow, produced by the sequence percolation – water table – lateral flow to streams, is lost. These surges in flow may disrupt stream ecosystems and if discharge is very large compared with stream capacity, flooding may result. The effect of development, even on relatively small sites, can be to prevent recharge of groundwater. Under such conditions water tables exacerbated by abstraction may fall, particularly when they cannot be recharged by lateral movement from the surrounding area. This may have profound effects on the ecology of a development site, which may reduce the effectiveness of landscape and site conservation schemes. This situation may be aggravated if development involves extensive trenching for drains or pipelines as these may act as effective land drains on-site. Drainage of a site by this means may be particularly significant if any habitats within the immediate area are dependent upon a high water table; examples are marshes, bogs, fens and swamps. Not only may development prevent the recharging of groundwater on which such communities may be dependent, but active drainage may rapidly destroy much of their interest (Canter 1991a).

Another important consideration relating to groundwater is the existence of aquifers, which are used as a source of supply. Agricultural chemicals, particularly nitrates, have shown increased levels in groundwater from percolation through the soil. In project appraisal attention should be focused on risk of contamination with toxic chemicals. The problems in confined and unconfined aquifer contamination are different. If two developments contaminate groundwater, normal monitoring procedures may detect increased levels in the river relatively quickly whereas contamination of the confined aquifer might only be detected when a large part of it had become unserviceable. Any possible groundwater contamination should be identified at an early stage in project appraisal. An activity that would need special consideration is leakage of toxic materials from storage vessels or treatment areas for toxic products including dumps, tailings lagoons and settling ponds. Any geological faulting in the area, which may lead to leakage, should be considered.

When undertaking a groundwater impact study the following steps may be followed (Canter 1991a):

◆ *Step 1: Determine water quality/water quantity impacts from proposed project/plan*: Identify the type and quantities of water pollutants and/or water quantity changes anticipated.

◆ *Step 2: Describe existing groundwater resource conditions*: Describe existing groundwater quality and quantity characteristics for the planning area for the project.

◆ *Step 3: Describe unique groundwater quality/quantity problems*: Identify areas which exhibit special or unique problems that should be addressed as part of describing the baseline conditions for the groundwater resources in the study area.

◆ *Step 4: Identify applicable groundwater standards*: Outline pertinent groundwater quality/quantity standards for given geographical areas.

◆ *Step 5: Document existing or potential sources of groundwater pollution/ groundwater usage in the study area*: Identify other potential and actual sources of groundwater pollution already existing in the study area.

◆ *Step 6: Consider phase impacts*: Consider component parts of the project (e.g. construction) and their associated impacts.

◆ *Step 7: Determine mesoscale impacts*: Consider large-scale potential impacts of the proposed project (e.g. total groundwater system). In order to achieve Step 7, it is necessary to have delineated the types and quantities of water pollutants and/or water quantity requirements of the proposed project (Step 1), and also to have determined existing conditions in terms of groundwater pollution sources and groundwater usage (Step 5).

◆ *Step 8: Determine microscale impacts*: Microscale impacts refer to the small scale; specifically, this relates to the potential impacts of the proposed project on the local groundwater within the project boundary.

◆ *Step 9: Consider mitigation or control measures*: Consider ways in which negative impacts could be minimised.

◆ *Step 10: Consider other related impact issues*: Consider where appropriate.

Both surface water and groundwater if used for a specific purpose will require an acceptable quality prior to its use. Acceptable criteria are detailed in the National Society for Clean Air and Environmental Protection Handbook (1998). A number of European Union (EU) Directives define the acceptable quality of water for particular purposes and make provision for both achieving and monitoring the quality of water. Directives have been adopted concerning the quality of surface water intended for the abstraction of drinking water, for bathing water, freshwater fish, shell fish and for water intended for human consumption.

The EU controls on water pollution can be classified into three categories: discharge of dangerous substances (e.g. black and grey list substances (Framework Directive (76/464/EEC)); quality objectives (e.g. surface water (75/440/EEC), bathing water (76/160/EEC), drinking water (80/778), etc.) and by sector or industry (e.g. the titanium oxide industry). Discharges to groundwater are dealt with by Directive 80/68/EEC.

CASE STUDY: URBAN STORMWATER RUNOFF ASSESSMENT

Stormwater runoff and receiving stream water quality was monitored over a 20-month period for a 243 ha separately sewered housing estate to the north-west of London, UK. A description of the site and instrumentation is given elsewhere (Harrop 1984). The quality of the receiving stream was of special interest as it was a tributary of a river used for recreational purposes. The monitoring of surface runoff to the stream, and the identification of the pollutants carried with it, was therefore of particular importance. Water samples were taken upstream and downstream of the stormwater outlet pipe.

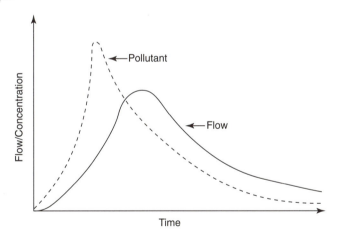

FIGURE 3.4 First flush effect of stormwater discharges

Pollutant concentrations on average were greatest downstream of the stormwater outlet; the reduction in pollutant levels compared to concentrations in stormwater runoff suggested that there was some form of filtering, storing and/or diluting of pollutants within the stream. Pollutant concentrations after stormwater discharge showed that on average surface runoff contaminants caused increases of up to 97 per cent in pollutant concentrations downstream. Peak parameter levels were between two and six times the acceptable level for potable water abstraction. Previous studies (Ellis 1976, Mance and Harman 1978 and Tucker 1975) have observed an initial peak flushing of contaminant levels prior to maximum flow on the discharge hydrograph. Studies showed that, with the exception of suspended solids, pollutants exhibited a marked first flush pattern. Suspended solids showed a more even distribution of maximum concentration on both the rising and recessional limb of the discharge hydrograph. Research has shown that the first flush of some individual pollutants occurs 20–25 minutes prior to the peak flow (Figure 3.4). This usually occurred when the hydrograph rising limb was protracted before peak flow was achieved. Nitrates, NH_4^+ and PO_4 showed pronounced first flushing which may partially be attributed to the rapid delivery of nutrients from the roadside gully-pot chambers during the initial rainfall-runoff process. The initial discharge of nutrients from gully-pots has previously been observed by Fletcher *et al.* (1978). The percentage first flushing occurrence of NO_3^-, NH_4^+ and PO_4 was 62.5 per cent, 80 per cent and 100 per cent respectively. A similar pattern of 74 per cent was noted for dissolved solids. The first flush occurrence of volatile dissolved solids (73 per cent) and volatile suspended solids (59 per cent) is very much in accordance with other studies (Wilkinson 1956), which found that initially stormwater solids were highly organic in nature. The first flush nature of stormwater discharges is in agreement with Ellis's (1984) work in the same catchment, with the occurrence being more frequent than that reported in the earlier work of Wilkinson (1956).

Stormwater surges can also modify the morphology of the stream itself (Gregory and Walling 1970 and Hollis and Luckett 1976). The receiving river showed signs of river bed scouring, resulting in a doubling of the river bed's width and shoaling of particulates and washed off debris.

Stormwater runoff enhances receiving stream water contaminant concentration, as well as influencing the morphology of the river bed. Runoff events showed pronounced first flushing in pollutant concentrations, with a series of storms reducing potential surface pollutant loadings. Peak pollutant concentrations were above those suitable for potable water abstraction.

Archaeological and cultural heritage assessment

Archaeological and cultural impact

For many countries there is provision for assessment of impacts on cultural heritage in the preparation of an ES. Experience suggests, however, that in most instances, the decision that is eventually reached is likely to be based on a restricted set of issues of major importance; and cultural matters on their own are unlikely to feature in this list – partly because other constraints (such as the minimisation of pollution) may be adjudged more significant and partly because – in cost terms – mitigation measures for aspects of the cultural heritage may be low in relation to the overall cost of the project (Ralston and Thomas 1993). A major difference between archaeological assessment and the assessments undertaken in many other disciplines that contribute to an ES is that the absence of evidence does not constitute evidence for absence. A further truism, shared with other protective measures, is that – unlike certain other resources for which subsequent recovery is a possibility – archaeological resources are finite and irreplaceable (Ralston and Thomas 1993).

Assessment methodology

Available archaeological assessment techniques vary from remote sensing through geophysical survey and field-walking to exploratory excavation. It would, however, appear to be not uncommon practice for the assessment to be restricted to a desk-bound procedure (e.g. review of maps, historical records, etc.) until the necessary planning consents have been obtained. For example, in the case of major infrastructure projects, such as pipelines or roads, etc., the land affected may not be in the ownership of the developer and therefore there may be problems with access. Equally, there is an understandable reluctance on the part of developers to invest heavily in activities involving potentially substantial costs, until such time as the development has been granted planning consent, at least in outline. An ES, however, would be expected to demonstrate familiarity with both the types and numbers of sites of all kinds within the proposed development area, as well as announcing appropriate strategies for

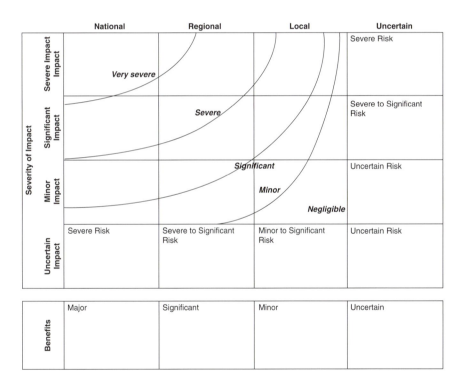

FIGURE 3.5 Archaeological assessment procedure to evaluate the severity of impact
Note: Matrix developed by the Oxford Archaeological Unit for the Channel Tunnel Rail Link Project to assist rigorous and consistent assessment of the severity of adverse effects. The numbers placed on the matrix represent affected features, their positions being determined by their importance and the scale of the predicted impact. (Copyright British Railways Board and Oxford Archaeological Unit)
Source: Oxford Archaeological Unit (1991)

enhancing the archaeological record for those portions of the development area for which there is reason to believe that the archaeological record, assembled on the basis of publicly accessible information, is incomplete or otherwise defective. If required, field work should be programmed as early as possible in the assessment process (Ralston and Thomas 1993).

With regard to the identification of physical features of the historic environment, data gathering can be seen in terms of four typical stages (Lambrick 1993): consultation of statutory lists, registers, and records; detailed research of original secondary documentation; general 'walkover' survey; or detailed field and structural surveys, where necessary staged in terms of general perception and survey, and detailed evaluation. Each of these stages may contribute to site or route selection and subsequent stages in the more detailed development of the project. It is not possible to be dogmatic about how precisely these stages of research should match up with the development stages of the project since this will vary according to its type, scale and practical logistics. The key point is that EA should involve assessment of all the aspects of the historic environment to a level of detail commensurate with the requirements of sound decision-making, at each major phase in the planning of the particular project concerned.

CASE STUDY: ARCHAEOLOGICAL ASSESSMENT PROCEDURE

Figure 3.5 shows a matrix developed by the Oxford Archaeological Unit (Oxford Archaeological Unit 1991) for the Channel Tunnel Rail Link Project. This was used to help assess the severity of adverse archaeological impact.

Social impact assessment

Social impacts

Social impact assessment (SIA) can be defined as the process of assessing or estimating, in advance, the social consequences that are likely to follow from specific policy actions or project development, particularly in the context of appropriate national, state or provincial environmental policy legislation. Social impacts include all social and cultural consequences to human populations of any public or private actions that alter the ways in which people live, work, play, relate to one another, organise to meet their needs, and generally cope as members of society. Cultural impacts involve changes to the norms, values, and beliefs of individuals that guide and rationalise their cognition of themselves and their society (Burdge and Vanclay 1996).

Assessment methodology

The ultimate objective of SIA is to evaluate the potential social impacts to be expected from project development. Such an assessment will contain specific recommendations for modification of project siting, design or development and proposals for mitigation, which will avoid, minimise or counterbalance potential impacts. SIA encompasses a wide range of social and economic concerns, some of which are listed under the six basic headings given below (Westman 1985).

Health and safety

◆ Crime levels.
◆ Public risk of injury, health impairment, or death.
◆ Psychological environment, including anxiety levels, personal comfort and enjoyment, privacy.

Economy

◆ Employment.
◆ Housing.
◆ Commerce.
◆ Cost of living.

Cultural and urban resources

◆ Belief systems: religious, political, social values.
◆ Identification: nationality, indigenous, ethnic, racial, life style, community.
◆ Recreational and scientific resources.
◆ Historical and archaeological resources.

Aesthetic characteristics

◆ Odours.
◆ Noise.
◆ Atmospheric visibility.
◆ Visual qualities of landscape.
◆ Vibrations.
◆ Light quality and quantity.

Regional growth and infrastructure

◆ Provision of social services, including police, fire protection, energy, sewage treatment, water supply, flood protection, solid waste disposal, health care, transport, education.
◆ Changes in land use.
◆ Government laws and policies, including zoning policies, national and local plans, environment and land use laws and policies, antiquities and historical preservation acts, treaties and other governmental obligations.

Population characteristics

◆ Birth and death rates.
◆ Density and distribution.
◆ Immigration and emigration.
◆ Age structure.
◆ Sex ratio.

However, Burdge and Vanclay (1996) have identified several problems confronting SIA in the form of applying social sciences to SIA; problems and difficulties with the SIA process itself; problems with the procedures applying to SIA and a prevailing 'asocietal mentality' – an attitude that humans 'don't count' by the commissioners of SIA.

Among the most controversial types of development project, in terms of their range and magnitude of social problems, are large hydroelectric dam proposals. The biggest failures have involved inadequate resettlement plans. Goldsmith and Hildyard (1984) provide detailed documentation of problems, such as inadequate compensation for land, ethnic differences being ignored, inappropriate housing, and health problems. A subsequent publication (Goldsmith and Hildyard 1986) gives 31 case studies from temperate and tropical regions. Further examples are cited in Castro-Morales and Gorzula (1986).

In any large-scale study there will always be limitations with regard to the amounts of money, time and personnel that are available and big projects give little flexibility to allow for wasted time and mistakes. Therefore, efforts must be focused and co-ordinated. The basic methods involve:

◆ gathering information on existing conditions;
◆ designing and conducting special studies to develop any additional necessary data; and
◆ integrating and analysing the resulting information and data.

The organisation of the study is, however, contingent on the type of project, its magnitude and the extent of the expected impacts. Burdge and Vanclay (1996) have established that the SIA process provides direction in:

◆ understanding, managing and controlling change;
◆ predicting probable impacts from change strategies or development projects;
◆ identifying, developing and implementing mitigation strategies in order to minimise impacts;
◆ developing and implementing monitoring programmes to identify unanticipated impacts that cause social change;
◆ developing and implementing mitigation mechanisms to deal with unexpected impacts as they develop; and
◆ evaluating impacts caused by earlier developments, activities, etc.

In theory, social impacts should be assessed during the three project phases (planning, construction and operation). In practice, many such studies are not contemplated until planning is well under way and the general characteristics of the development have already been established. This is particularly true in large-scale and long-term projects such as Guri, Venezuela, where the feasibility studies were carried out and the major policy and programme decisions were taken during the 1950s. Thus, in the case of the Guri dam the impacts on the economy, job creation, regional urban development and demographics were not considered as part of the SIAs carried out during the 1970s and 1980s because they were already the prime objectives of the regional development programme.

CASE STUDY: GURI HYDROELECTRIC POWER PROJECT, VENEZUELA (GORZULA 1992)

A detailed overview of the significance of EA and the Guri hydroelectric power project is given in Castro-Morales and Gorzula (1986). The social impact of dams is enormous during the construction phase. It is stressed, however, that in this type of development environmental impact must be seen as a two-sided coin. The potential impact of the environment upon the project can be just as important as the impact of the project upon the environment. During the operational phase human activities can have direct effects on the scheme's functioning, efficiency and life-span.

In the past many operational problems for large dams have been caused by the uncontrolled settlement of the lands adjacent to the artificial lake. Attracted by the promise of abundant and free water supplies for irrigation, uncontrolled settlement can result in the eutrophication and the accelerated sedimentation of the impoundment. From 1963 to 1986 the Guri project required an investment of US$ 5 billion. In order to help protect this investment it was decided that, in addition to relocating the inhabitants from the area to be flooded, there should be the establishment of a protected area up to the watershed surrounding the impoundment. The first problem was to discourage any immigration of additional people into the area. This was achieved during 1979 by making an inventory of every single inhabited structure, and marking it with an identification number that could be seen from a helicopter. It should be stressed that immigration per se was not prohibited. It was simply pointed out to any new people coming into the area that they would receive no compensation whatsoever once the time for relocation came. The result was that only sporadic immigration occurred, and this was monitored and controlled with the help of the National Guard. The total number of persons who were to be directly affected by the flooding was close to 4,185. Of this total 22 per cent were represented by dispersed inhabitants, and the remaining 78 per cent were concentrated in three villages, three hamlets and a low-lying quarter of a small town. The dispersed inhabitants fell into two basic categories: those with title deeds to their property and those without.

In general, the people who possessed title deeds were ranchers whose properties were easily quantifiable in terms of area, because they were fenced and registered.

Such ranches had measurable infrastructure in terms of the size and the type of dwelling-houses and out-buildings, size and type of electrical generator, kilometres of fencing, number of wells and bore-holes, number of artificial ponds, and hectares of pasture and crops, etc. Thus, the valuation of a large number of properties requires detailed planning and organisation. The people without title deeds were subsistence farmers. They occupied rather vague areas of land, had only rudimentary infrastructure and their use of the land could not be valued by any standard method. In this case a simple empirical approach was adopted. If, for example, the farmer showed that he used 30 hectares of land then he was paid compensation as if he were the owner of 30 hectares. If the family dwelling was a three-room wattle and daub structure with a palm thatch roof, the compensation was paid for the value of a similar sized building made out of bricks, concrete and a corrugated galvanised iron roof. The relocation of this group was successful, with the average compensation paid to each family being about US$ 3,500.

Of the seven settlements, four opted for cash compensation. The remaining three communities, totalling about 1,000 persons, chose the option of resettlement in a new village. From the outset it was decided that these people would play an active role in the design of their new homes.

The first community was an indigenous community consisting of almost 500 Pemon Indians. These people were living in very poor conditions. Their dwellings were made of mud, with earth floors and palm thatch or corrugated galvanised iron roofs. Water for domestic use was taken directly from the river. They had no sewerage system whatsoever and excrement was deposited in the open air. The villagers received no medical care, unless they attended the nearest clinic that was two hours' boat journey away. Many individuals showed symptoms of parasitosis and anaemia associated with malnutrition. Their productivity was based upon slash and burn agriculture, fishing and hunting. Apart from the annual sale of a few hundred tortoises (a traditional regional dish at Easter) they had little access to local markets.

A series of studies was initiated in order to characterise their epidemiology, nutrition, genetics, medical, social, agricultural, fishing and hunting activities. The inhabitants collaborated and provided EDELCA (Electrificación del Caroni, C.A.) staff with a dwelling for their own use. In the course of carrying out such studies valuable feedback was gained from the community, and anxiety about the inevitable resettlement was greatly reduced.

This indigenous community was moved to higher ground less than 1 km away. The new village with 75 houses for 418 people, a school house, a dispensary and a community centre was constructed within a cleared area of 50 ha. An electrical plant provided lighting for the streets and the houses. A sewerage system transported the effluent from the lavatories to an oxidation lagoon prior to discharge downstream from the village. Water for domestic use was pumped in from a floating intake upstream from the village, passed through a water treatment plant, and then to an elevated storage tank before being piped to each home. The villagers established several design characteristics that were compatible with the social structure of this ethnic group. For example, individual lavatories were placed away from the dwellings and potable water was available from taps outside the dwellings. The dispensary was supplied with the necessary furniture, medical equipment and medicines. Several members of the community completed courses in nursing and basic medicine, with the result that basic medical assistance is permanently available at the dispensary.

In the case of the other two communities, one necessitated the relocation of 50 families to a new part of the small town that they already inhabited. The other community consisted of a village that was developed in the 1940s during a diamond mining boom and had become a permanent agricultural settlement. The village had no running water, no sewerage system, no school and an electrical plant that only functioned for a few hours each day. The villagers' greatest worry was the relocation of their cemetery. The new village covered 12 ha. It had 75 dwellings for 375 people, an aqueduct, a sewerage system, a new electrical plant with street lighting, a furnished dispensary, a fully equipped school, a community centre, a plaza planted with trees and grass, a children's park and a new cemetery which contained graves with all of the corpses from the old cemetery. The new village was fully covered by the national policy of consolidation of rural communities, which suffer from population loss due to urban migration. The inhabitants were therefore eligible for a wide range of agro credits and technical assistance. The 10-year programme for relocating the three communities cost in the order of 47 million bolivares (EDELCA 1987) .

Questions for thought

1. What are the key component parts of applying an EA technique?
2. How would you apply the techniques detailed in the chapter to:
   ◆ an industrial manufacturing complex (e.g. petrochemical plant, smelter);
   ◆ a tourism resort (e.g. golf course, hotel and water sports complex);
   ◆ a waste management facility (e.g. landfill, incinerator, waste transfer centre)?
3. How can the EA practitioner reconcile the use of both qualitative and quantitative assessment in the EA process?

Chapter 4

Environmental
risk assessment

◆ **Introduction** 74
◆ **Terminology** 75
◆ **Applications of risk assessment** 79
◆ **Questions for thought** 88

Introduction

Public concerns and anxieties have focused attention on the need for informed and transparent assessment procedures to ensure that the impact of developments is acceptable from an environmental perspective. Therefore, it is essential that due attention is given to environmental risk assessment procedures as well as EA for certain projects. Risk assessment is widely regarded as a specialised tool for high-profile or high-risk projects that require an examination of potential risk extending beyond a conventional source–pathway–receptor analysis. Recent guidance has been issued by the UK Department of Environment on risk assessment procedures for environmental protection (Department of the Environment UK 1995). The output of risk assessment and EA studies has become inextricably linked with the common objective of assessing the significance of impact. A growing number of practitioners are incorporating the procedures of risk assessment into EA studies.

Risk combines the probability of an adverse event (a hazard) occurring, with an analysis of the severity of the subsequent consequences. 'Risk assessment' can be simply defined as the process of assimilating and analysing all the available scientific information associated with a hazard or set of hazards (Rodricks 1992a). Pollard *et al.* (1995) have previously discussed a number of approaches to risk assessment applied within environmental management and these can be considered in a general sense as being qualitative, semi-quantitative or quantitative in nature.

Environmental risks will always be a contingent of the variable conditions of exposure and, under a range of exposure scenario conditions, risk can be envisaged as a distribution of risks ranging from low to high-risk conditions. High-risk situations are those where direct short exposure pathways exist from a high hazard source to a particularly sensitive receptor; whereas low-risk situations are those where indirect or diffuse pathways exist from a source of low hazard to a receptor of low sensitivity, or where substantial dilution occurs along the exposure pathway (Pollard *et al.* 1995).

As a management tool, quantitative risk assessment (QRA) attempts to express risks mathematically by modelling exposures to source(s) present, aggregating exposures over all relevant exposure routes and key sources, and expressing estimated risks for individual receptors or groups of receptors. Throughout preparation and communication of a QRA, it is essential to keep in focus the overall objective of the risk assessment in the context of the EA to avoid misuse of output. The objective of the QRA is to evaluate the potential incremental adverse effects that could arise from a development under the specific conditions of the exposure scenario. This is usually performed with reference to communities (or other receptors) downwind of the facility at the point of maximum ground level pollutant concentrations (Harrop and Pollard 1998). In essence, the practice of QRA in the UK at present is one largely based on US experience and implementation.

Terminology

Risk assessment is a process in which the probability or frequency of harm for a given hazard (an event or agent which has the potential to cause harm) is estimated. Petts (1994) and Pollard *et al.* (1995) identify four principal stages of risk assessment (Figure 4.1 summarises these procedural steps):

◆ identification of the sources and components of hazards at a facility;
◆ analysis of the fate of the hazardous substances in the environmental media through which they are transported; determination of the release probabilities, quantities, and rates; identification of exposure pathways by which substances could reach receptors, and the sensitivity characteristics of the receptors at risk;
◆ estimation of risk with reference to an accepted dose–response relationship; and
◆ evaluation as to the acceptability, or tolerability, of the estimated risk.

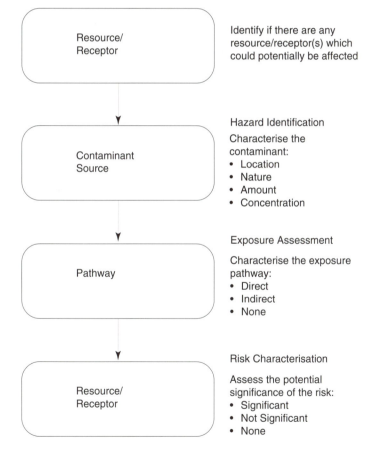

FIGURE 4.1   QRA procedural steps
Source: Pollard *et al.* (1995), CIWEM

Within environmental risk assessment, these stages refer to the identification of chemical or physical agents posing adverse health effects, an analysis of the amount of exposure received by the receptor(s), an estimation of the relationship between exposure and toxic response, and finally, the evaluation of exposure estimates through their translation into integrated, and ideally probability-based, risk distributions. Several texts examine the fine detail (Rodricks 1992b and Gotts 1993) of risk assessment. Generally, risk assessment practices are determined by a combination of factors including scientific and technical knowledge, the level of experience of risk assessors, specific local conditions, community concerns and environmental regulations and guidance (Price 1994). The four stages of risk assessment are described next.

Hazard identification

Hazard identification involves identifying the chemicals/materials present; determining their quantity, form and location; and selecting the indicator chemicals (i.e. those that could contribute to the greatest proportion of the risk) (Petts and Eduljee 1994).

Hazard analysis

Hazard analysis is the process of determining the release probabilities, quantities and emission rates, the exposure pathways by which the released substance(s) could reach sensitive receptors, and the characteristics, particularly the differing sensitivities of the receptors at risk.

Risk estimation

Risk estimation involves application of the dose–response relationship between a receptor and a pollutant to estimate incremental risks for specific contaminants over the duration of exposure. Environmental exposures usually result in relatively low intakes. It is often assumed that the low dose–response relationship is linear in the low dose region of the dose–response curve. Under this assumption, the slope of the dose–response curve is equivalent to the carcinogenic potency factor, and the estimated risk will be related to intake at low exposure.

Risk evaluation

The Royal Commission on Environmental Pollution (RCEP) (Royal Commission on Environmental Pollution 1989) suggests that there is a broad consensus among regulatory authorities in the UK and the US that annual incremental risks of death greater than 1 in 10,000 are too high to be acceptable; and that a risk of 1 in a million represents a reasonable upper bound limit, beyond which measures to achieve a further reduction in the risk would not be justified in terms of the benefit gained. The Health and Safety Executive (HSE) has presented criteria (Health and Safety Executive 1989) designed to advise planning authorities on the development of land in the vicinity of 'major hazard' installations. Although not all developments may be classified as a 'major hazard' installation, the criteria for acceptable risk suggested by the HSE form the basis of a criterion for many studies. The HSE proposes that housing developments providing for more than about 75 people should be subjected to a calculated individual risk of less than 1 in a million per year. The same value is recommended by the RCEP.

Table 4.1 shows some of the risks arising from exposure to activities and events encountered in daily life as a risk per million per year. In comparison, Table 4.2 shows the calculated risks for proposed incinerator studies (Harrop and Pollard 1998).

As with any mathematical estimation model (see Chapter 3), QRA has its limitations and these invariably act to qualify the estimates of risk and place constraints on the means and manner in which the risk can be estimated, expressed, communicated and ultimately managed. Because any risk assessment requires professional value judgements to be made regarding data input, it must be realised at the outset that there remains an element of subjectivity regarding the output. This is particularly the case where emissions and incremental exposures from proposed developments have to be modelled as opposed to being measured at an operational facility (Harrop and Pollard 1998).

Harrop and Pollard (1998) have previously considered the question of whether QRA is appropriate for EA. Considering the constraints on obtaining site-specific data for proposed developments and the largely unresolvable limitations of the baseline toxicology, it is entirely valid that the use of QRA is reviewed critically for EA

TABLE 4.1   Risk assessment criteria

Causes	Risk per million per year
All causes (mainly illnesses from natural causes)	11,900
Cancer	2,800
(These figures vary greatly with age)	
All violent causes (accident, homicide, suicide etc.)	396
Road accidents	100
Accidents in private homes (average for occupants only)	93
Fire or flame (all types)	15
Drowning	6
Gas incident (fire, explosion or carbon monoxide poisoning)	1.8
Excessive cold	8
Lightning	0.1

Source: Environmental Resource Management (ERM) (1993)

TABLE 4.2 QRA predictions for proposed incineration plants in the UK

Type of incinerator	Level of annual risk (range where available)
Waste-to-energy (WtE) plant	1 in 33,000,000 to 55,000,000
	1 in 26,000,000 to 26,900,000
	1 in 15,900,000
Sewage sludge	1 in 2,400,000 to 5,500,000
	1 in 12,500,000 to 58,000,000
Clinical waste	1 in 9,000,000 to 38,900,000
Industrial waste	1 in 7,800,000 to 12,100,000

Note: The values given in TABLE 4.2 have been taken from a series of ESs undertaken by ERM and Aspinwall & Company. Because the detail of assessment undertaken for each study was variable (e.g. inhalation exposure, ingestion or total exposure assessments, realistic or worst case scenarios, etc.) direct comparisons between studies should be avoided. The values given are indicative of possible exposure levels and are not absolute. For specific details the respective ESs should be carefully consulted.
Source: Harrop and Pollard (1998)

applications. The underlying premise of QRA is that very low environmental exposures can be mathematically translated into expressions of the probability of adverse effects occurring and their associated significance. It is usual to assess both potential carcinogenic and non-carcinogenic effects (so-called toxic 'end-points'). This is conventionally undertaken by the use of excess lifetime cancer risks and hazard quotients, respectively. Comparison of the risk estimates and hazard quotients obtained with reference to accepted tolerability criteria and background risk levels provides an account of the incremental risk and its significance attributable to the proposed facility. The assessment of carcinogenic and non-carcinogenic risk relies fundamentally and heavily on interpretation of the baseline toxicology. Because much of the data on chemical exposure has been gathered on experimental animals (rather than humans) at high doses (rather than low), and at doses deliberately sufficient to solicit a toxic response, there is an over-riding need for extreme caution in the interpretation of QRA output. Hence risk assessment output is not absolute and does not represent the actual probability, say, of contracting cancer. The output is, instead, indicative and is used to highlight the relative distribution of risk among exposure pathways and between individual receptors so that risk management measures, where required, can be prioritised and designed accordingly. Understanding this is essential if the practitioner is to avoid over-interpretation of output and an assumption that the risk estimates produced are in some manner more precise than the fundamental science on which they rest.

Some authors have argued that QRA methodologies are inappropriate because they cannot resolve the considerable uncertainties associated with establishing safe levels of exposure for genotoxic carcinogens (Maynard *et al.* 1995). Application of low-dose extrapolation to derive a *de minimus* risk level has defensibility problems (Hrudey and Krewski 1995 and Kelly 1991) and has not been entirely successful in yielding 'safe' levels of exposure to carcinogens. In the long term, there should be investment in the powers of environmental epidemiology to provide direct estimates of risk on human receptors, but such approaches provide only retrospective views, are complicated by high background rates of cancer and are still limited at very low levels of exposure by the practical and ethical constraints of studying human subjects (Doll 1995 and Coggon 1995).

Applications of risk assessment

Pollard *et al.* (1995) have previously applied risk assessment methodology to potentially contaminated sites along the route of a railway link. The study involved the assessment of 228 potentially contaminated sites. The screening of risks was effectively streamlined through the use of a qualitative assessment matrix for each receptor type (e.g. for groundwater acting as a receptor) (Table 4.3). For risk characterisation, sensitivity criteria were established prior to individual screening through professional experience and judgement. Hazard identification was performed by relating historical operations and site management practices obtained via a desk study and site walk-over to potential contaminative land use, and hence to contaminant characterisation. Effects were classified in relation to the nature and quantity of possible source contaminants present on the basis of site use, professional judgement and established guidance. Exposure assessment was performed by classifying pathways as direct (e.g. contact), indirect (e.g. via another medium such as inhalation of windblown dust) or none. Finally risk characterisation was achieved by assessing the significance of source pathway and receptor and the sensitivity of the receptor understudy (Pollard *et al.* 1995).

Morrey *et al.* (1996) attempted to quantify the level of risk associated with a composite lined landfill with a low permeability cap. Although the leachate was designed to be contained, it was recognised that there was still a small risk of leakage from the site. Concern was expressed over the potential risk to public groundwater supply 1 km from the site. The study considered the following factors as an analysis of risk: the quantity and quality of leachate which may pass through the liner and the identification of the potential pathways which any leakage leaving the site may follow, the water supply sources which may be affected and the magnitude of any effects. The assessment was made to determine whether there would be any groundwater pollution. The criteria used were the European drinking water quality standards. It was considered that a reasonable worst case leakage situation would be within the landfill restored, capped but with a clogged drain blanket. The expected (p = 0.5) leakage was 8.5 litres per hectare per day (l ha^{-1} d^{-1}) while the 90th percentile (p = 0.9) leakage was 17.6 l ha^{-1} d^{-1}.

TABLE 4.3 Qualitative assessment matrix for groundwater receptors

	Source		Pathway	Effect on receptor	
Nature	Concentration	Quantity		Sensitive	Less sensitive
		Major	Direct	S	NS
			Indirect	S	NS
	High		None	N	N
		Minor	Direct	S	NS
			Indirect	S	NS
			None	N	N
Likely to be a		Major	Direct	S	NS
potential risk			Indirect	S	NS
	Intermediate		None	N	N
		Minor	Direct	S	NS
			Indirect	S	NS
			None	N	N
		Major	Direct	S	NS
			Indirect	S	NS
	Low		None	N	N
		Minor	Direct	S	NS
			Indirect	S	NS
			None	N	N

S – potentially significant
NS – no significant effect
N – no effect
Source: Pollard *et al.* (1995)

By applying a simple mixing model Morrey *et al.* (1996) demonstrated that each of the above leachate levels resulted in no breach of the European water quality standards at the public water supply borehole, even in the absence of any attentuation. The risk assessment together with planned contingency measures was used to satisfy a UK Inspector at a Public Inquiry that the proposed landfill operation did not represent an unacceptable risk of groundwater pollution.

CASE STUDY: QRA OF A WASTE TO ENERGY PLANT

To date QRA has been used in a limited, although growing, number of EAs in the UK. One area of EA where QRA has been applied is in the assessment of proposed incinerators.

Incineration and potential risks to public health are often inextricably linked in the public mind. It is paramount, therefore, that due consideration is given to assessing the effects of these facilities on the public in an informed and reliable

TABLE 4.4   Maximum long-term annual ground-level concentrations and recommended air-quality guidelines for a proposed waste to energy plant ($\mu g\ m^{-3}$)

Pollutant	Average measured baseline concentration	Predicted maximum increase in ground level concentrations	Guideline OEL/100*
VOC	144	0.04	—
CO	2,900	0.63	550
SO$_2$	14–28	0.08	50
NO$_2$	24.0–44.2	2.36	50
HCl	1.56	0.24	70
HF	0.95	0.0079	25
Cd	0.001	0.0004	0.25 (0.05)**
Hg	0.006	0.0004	0.5
As	0.007	0.000068	1.0 (0.2)**
Pb	0.2	0.000034	1.5 (0.3)**
Cr	0.003	0.000034	0.5 (0.1)**
Mn	0.009	0.000034	10
Ni	0.005	0.000514	1.0 (0.2)**
Cu	0.03	0.000068	2.0
Sn	—	0.000034	20
Dioxins and furans***	0.24	7.9 x 10^{-4}	—
TSP	35.3	0.08	50

*   Occupational Exposure Limit (OEL) value divided by 100
**   Maximum Exposure Limit (MEL) value divided by 500
***   pg TEQ m^{-3}
—   no data

manner. The risk assessment process is unfortunately subject to considerable debate in the UK where a more pragmatic approach to environmental regulation has prevailed to date. Nevertheless, the practical use of currently available assessment techniques is gaining wider use and as available information on the methodologies develops then the procedures will improve. In the absence of epidemiological evidence and in the interests of public health protection, it is arguably prudent to apply informed QRA methods to assist in devising risk management strategies for stack emissions from incinerators.

A study looked at the emissions from a waste-to-energy plant (Aspinwall & Company 1994a). Emissions met Her Majesty's Inspectorate of Pollution (HMIP) Chief Inspector's Guidance to Inspectors, Process Guidance Note IPR 5/3 (Department of the Environment UK 1992a). Table 4.4 summarises the long-term predicted ground-level concentrations from the study.

Table 4.4 shows that the predicted increases in maximum long-term (annual) ground level concentrations were low compared with existing baseline air quality levels. However, the assessment did not allow for an adequate assessment of carcinogenic emissions without the application of QRA. The equations used for the calculation of unit carcinogenic risk by inhalation are as follows (United States Environmental Protection Agency 1986):

Risk = cancer potency factor × inhalation exposure $\qquad$ (1)

where:

inhalation exposure = (dose) ÷ (body weight (kg) × lifetime(days)) $\qquad$ (2)

dose (mg) = contaminant concentration (mg m^{-3}) × inhalation rate
(m^3 day^{-1}) × duration of exposure (days) $\qquad$ (3)

Input data and assumptions for equations (1) to (3) are given in Table 4.5.
It is customary in risk analysis to assume a worst case scenario so as to be protective of human health. Therefore it was assumed that emissions of the above contaminants were continuous over 365 days per annum for 70 years, the average lifetime of a hypothetical exposed individual. Further, the individual is assumed to have spent an entire lifetime at the point of maximum impact inhaling the emissions from the incinerator and absorbing 100 per cent of the material inhaled.

The total excess cancer risk for the above contaminants may be gained from the following equation:

Total Excess Cancer Risk = Risk 1 + Risk 2 + Risk 3 + ... Risk n $\qquad$ (4)

where:

Risk 1 = Individual excess cancer risk from a lifetime exposure to the first substance;

Risk n = Individual risk of additional substances.

Maximum annual predicted ground level concentrations of the carcinogenic contaminants (μg m^{-3}) are given in Table 4.4. The worst case scenario was used to predict the increased individual lifetime risk of developing cancer, to the hypothetical maximum exposed individual, as a result of inhalation of the emissions. The results, based on equations (1) to (3), are given in Table 4.6.

The predictions in Table 4.6 were performed on the assumption that the exposed individual was located out of doors for 100 per cent of the time; 100 per cent w/w of the inhaled material is absorbed into the lungs; and the plant operates for 8,760 hours (continuously all year). The cumulative effect of these factors is that the estimate of risk is multiplied by 0.7 (equivalent to approximately 17 hours a day outside) to obtain a level of increased lifetime risk, reflecting a more realistic worst case. Similarly, the total alveolar absorption of the above contaminants ranges from 0.25 for Cr to 0.75 for Ni. Therefore the estimate of risk multiplied by 0.525 (0.7 × 0.75) may give a more realistic value of 1.2×10^{-6}. Averaged over a typical lifetime of 70 years this risk corresponds to a value of 0.2×10^{-7} per year.

TABLE 4.5 QRA data assumptions

AT (Average human lifetime)	70 years
BW (Average adult body weight)	70 kilogrammes (kg)
Volume of inhaled air daily	20 cubic metres per person per day (m^{-3} person^{-1} day^{-1})
ED (Exposure duration)	70 years
Cancer potency factor	As = 5.0×10^1 milligrammes per kilogramme per day
	Cd = 6.1
	Cr VI = 4.1×10^1 (1% hexavalent chromium)
	Ni = 1.19
	Dioxins and furans =1.56×10^5 (2, 3, 7, 8 – TCDD)
C_{soil}	soil concentration ($\mu g\ g^{-1}$)
C_{air}	air concentration ($\mu g\ m^{-3}$)
D_{dry}	dry deposition velocity (m sec^{-1})
WO_{wet}	wet deposition velocity (m sec^{-1})
EF	exposure frequency (3.156×10^7 sec yr^{-1})
r_{soil}	soil bulk density (1.4×10^6 g m^{-3})
A_{soil}	soil volume exposed per unit soil area (5×10^{-2} m^3 m^{-2})
II_{soil}	exposure intake attributable to incidental ingestion of surface contaminated soil (mg kg^{-1} d^{-1})
IR_{soil}	ingestion rate of soil (mg soil d^{-1})
CF	conversion factor (10^{-6} kg mg^{-1})
B_{soil}	fractional bio-availability of contaminant *in vivo* (unitless)
EF	exposure frequency (365 d yr^{-1})
IR_{soil}	incremental carcinogenic risk attributable to soil ingestion (unitless)
$q_1{}^*$	chemical specific cancer potency factor; a plausible upper-bound estimate of the probability of a carcinogenic response per unit intake over a lifetime; here for oral exposure routes (mg^{-1} kg d).
DI_{soil}	chemical intake attributable to dermal absorption of surface contaminated soil (mg kg^{-1} d^{-1})
SA	skin surface area available for dermal contact (m^2 absorption event^{-1})
AF	soil to skin adherence factor (g m^{-2})
ABS	absorption factor (unitless)
EF	exposure frequency (365 d yr^{-1})
C_{food}	chemical concentration in food estimated from modelled C_{soil} values using literature plant–soil ratios (mg kg^{-1})
FI_{veg}	chemical intake attributable to home-grown vegetable consumption (mg kg^{-1} d^{-1})
IR	ingestion rate (kg d^{-1})
FI	fraction ingested (unitless)

Note: Inhalation rate of 20m^3 day^{-1} for a 70 kg person based on 'Health and Safety Component of Environmental Impact Assessment', World Health Organisation (World Health Organisation 1987b) and *Superfund Public Health Evaluation Manual* (United States Environmental Protection Agency 1986). The contaminants listed above were those identified as having carcinogenic properties in the study

TABLE 4.6 Predicted lifetime risk of developing cancer

	Increased lifetime risk – conservative estimates
As	9.7×10^{-7}
Cd	7.0×10^{-7}
Cr VI	4.0×10^{-7}
Ni	1.7×10^{-7}
Dioxins and furans	3.5×10^{-8}
Total	$\mathbf{2.3 \times 10^{-6}}$

The calculated annual aggregate individual risks for the conservative upper boundary and more realistic estimate fell well within the HSE criterion of 1 in a million per year. The calculated risk range of 0.2×10^{-7} per year corresponded to risks of 0.02 per million per year. For comparative purposes Table 4.1 shows the voluntary and involuntary risks arising from exposure to activities and events encountered in daily life as the risk per million per year.

The maximum long-term ground-level air concentrations of potential contaminants (Table 4.4) emanating from the proposed plant were considered as the potential sources of incremental soil contamination at the area of impact and were used to estimate the potential incremental carcinogenic risks and chronic non-cancer hazards attributable to indirect exposures. Translation of air concentrations to estimated soil concentrations from deposition was made by considering dry and wet deposition mechanisms for contaminants at the soil surface. Corresponding soil equivalent concentrations (C_{soil}) were estimated using equation (5) which averages the incremental concentrations delivered to surface soil over a 70-year period (see Table 4.5 for explanation of equation parameters):

$$C_{soil} = \{C_{air} \times (D_{dry} + WO_{wet}) \times EF \times ED\} / \{AT \times r_{soil} \times A_{soil}\} \qquad (5)$$

A value for dry deposition of 0.015 m s^{-1} was used, with wet deposition being estimated from a washout factor of 10^5 and a value for annual rainfall of 600 mm yr^{-1}. Soil bulk density (inclusive of void space) was assumed to be 1.4×10^6 g m^{-3} (Brady 1990), representative of cropped loamy soil and the duration of soil exposure taken to be 70 years. A soil cubic volume of 5×10^{-2} m^3 m^{-2} was selected as representative of the top 5 cm of a unit area (1 m^2) of soil.

In this conservative approach, maximum incremental soil concentrations result from deposition of the maximum long-term ground-level air concentrations over a 70-year period and were averaged over the lifetime of the facility to yield annual values. The adult human receptor (70 kg) was conservatively assumed to have been exposed to annual concentrations over an exposure duration of 70 years. Incremental soil concentrations attributable to the facility alone (i.e. above existing background concentrations) and estimates using equation (5) are shown in Table 4.7. No leaching from the top 5 cm of the soil, surface run-off or chemical transformation of individual

TABLE 4.7 Estimated soil concentrations (μg g^{-1}) resulting from deposition of maximum predicted long-term ground-level air concentrations (μg m^{-3})

Pollutant	Average measured air concentration (μg m^{-3})	Predicted maximum increase in ground-level concentrations (annual) (μg m^{-3})	Estimated incremental soil concentration (annual) (mg kg^{-1})
Cd	0.001	0.0004	0.003056
Hg	0.006	0.0004	0.003056
As	0.007	0.000068	0.000521
Pb	0.2	0.000034	0.000261
Cr	0.003	0.000034	0.000261
Mn	0.009	0.000034	0.000261
Ni	0.005	0.000514	0.003939
Cu	0.03	0.000068	0.000521
Sn	—	0.000034	0.000261
Dioxins and furans (TEQ)	2.4×10^{-7}	7.9×10^{-10}	6.05×10^{-9}

pollutant species was incorporated into the calculations. The predicted maximum increase in dioxin and furan (TEQ) concentrations was more than a thousandfold smaller than the mean (n = 133) European soil concentration of 8.7×10^{-6} mg kg^{-1} (United States Environmental Protection Agency 1994) and was therefore regarded as negligible.

Consideration of additional exposure pathways associated indirectly with airborne emissions was made following an evaluation of the widely accepted source–pathway–receptor model applied in turn to each pathway. Receptors may be connected with the hazard under consideration via one of several exposure pathways. Without the three essential components of a source (hazard), pathway and receptor, there can be no probability of the hazard being realised and hence no risk. Thus, the mere presence of a hazard does not ensure de facto that there will be attending risks. For the facility under consideration, the following indirect exposure routes were initially screened according to the above risk assessment philosophy:

◆ incidental ingestion of contaminated surface soils;
◆ dermal absorption of contaminated surface soils;
◆ consumption of vegetable produce grown in contaminated surface soils;
◆ consumption of contaminated fish;
◆ ingestion of drinking water; and
◆ dermal contact with contaminated water.

On the basis of professional judgement, expected human behavioural habits in the vicinity of the proposed facility and the negligible probabilities of even reduced exposures from the latter three indirect pathways, fish, drinking water and contaminated water pathways were screened out from the assessment which concentrated on the first three pathways as potential contributors of significance to overall risk from the facility.

Incidental ingestion of contaminated surface soils

Indirect exposure pathways including incidental ingestion of contaminated surface soils, dermal contact with contaminated surface soils and ingestion of contaminated produce grown in contaminated surface soils may contribute to the overall incremental carcinogenic risk and were estimated using an accepted US EPA risk assessment methodology (United States Environmental Protection Agency 1994).

Interpretation of carcinogenic risks was made with the realisation that to be protective of public health, the models used to estimate human exposure are necessarily conservative and embody many conservative assumptions aggregated in the final risk estimate. Hence risk estimates were not assessments of the actual likelihood of contracting cancer from exposures attributable to the facility. They were, however, worst case estimates of the magnitude of potential risks that can be compared with risk acceptability criteria. Contributions to carcinogenic risk from incidental soil ingestion can be estimated using the accepted equation for residential exposure (United States Environmental Protection Agency 1994) (see Table 4.5 for explanation of equation parameters):

$$II_{soil} = \{C_{soil} \times IR_{soil} \times CF \times B_{soil} \times EF \times ED\} / \{BW \times AT\} \qquad (6)$$

For an adult receptor, US EPA (United States Environmental Protection Agency 1990) recommends 100 mg soil d^{-1} as an upper bound estimate of adult ingestion for soil and a value that may also be representative of a child with a high tendency to ingest. This value is a conservative estimate and protective of public health as evidenced by a review of child soil ingestion (Fergusson and Marsh 1993). Bio-availabilities for all species were assumed to be 1 (i.e. $100\%^{w}/w$) and represent complete *in vivo* desorption from bound particles and full toxicokinetic absorption to target organs or systems. In reality, only a fraction of contaminants that reach the body will be absorbed and, of that fraction, only a further portion will reach the target system or organ.

The worst case estimates for adult intakes of contaminated surface soil via incidental ingestion and corresponding unit carcinogenic risks estimated using equation (7) (see Table 4.5 for explanation of equation parameters) were individually aggregated to a total estimate of carcinogenic risk of 1.1×10^{-7} over a lifetime which represents an aggregated annual risk of 2×10^{-9}. This is well below HSE risk acceptability criteria of 10^{-6}. Realistic worst case estimates for soil ingestion were likely to be well below this order of magnitude when taking into account actual ingestion rates and the low bio-availabilities for most of the above species.

$$IR_{soil} = II_{soil} \times q_1^* \qquad (7)$$

Even accounting for the uncertainties associated with estimation of these intakes and their conversion into potential carcinogenic risks, contributions to increased cancer potential from this pathway were regarded as negligible.

Dermal absorption of contaminated surface soils

Contributions to incremental carcinogenic risk from topical (skin) contact with contaminated surface soils can be estimated using equation (8) (see Table 4.5 for explanation of equation parameters). Outdoor exposures only were considered, with the adult human receptor spending 100 per cent time outdoors exposed to contaminated surface soils, and this exposure averaged for carcinogenic risk over a 70-year lifetime.

$$DI_{soil} = \{C_{soil} \times CF \times SA \times AF \times ABS \times EF \times ED\} / \{BW \times AT\} \qquad (8)$$

A value for the area of exposed skin on an adult working outside (SA; 0.17 m^2) was provided by Hawley (1985) and a soil to skin adherence factor of 27.7 g m^{-2} used for outdoor adult exposure based on kaolin clay on hands (United States Environmental Protection Agency 1994). For the potential contaminants studied, only arsenic was regarded as a potential carcinogen via dermal exposure.

A value of 0.1 (10%^{w}/w absorption) was assumed for the absorption factor of arsenic (APCOA 1992) across the skin barrier and is representative of the more absorptive metals and organics (APCOA 1992). For dioxins and furans, a value of 0.03 (United States Environmental Protection Agency 1994) was applied. An exposure frequency of 365 d yr^{-1} was assumed and the entire exposure averaged over a 70-year lifetime to yield chemical specific increased lifetime risks. Potential carcinogenic risk contributions from this pathway were regarded as negligible on the basis of an estimated annual individual risk of approximately 10^{-11}.

Consumption of vegetable produce grown in contaminated surface soils

Additional contributions to potential carcinogenic risk via oral exposure may come from the consumption of home-grown produce grown in contaminated surface soils. Produce can potentially become contaminated through dry or wet deposition on to the edible leaves of plants or via translocation to edible plant parts through the root–stem system from the contaminated surface, or near surface soil horizon. As with all exposures, uptakes are chemical and exposure route specific. Heavily contaminated soils may merit an examination of foliar deposition and translocation mechanisms as individual pathways likely to present significant contributions to overall exposure. In this case, in consideration of the low incremental concentrations attributable to the facility, exposures by the food pathway were estimated using generic plant–soil ratios where available rather than route-specific uptakes for individual vegetables.

Residential exposures attributable to the ingestion of vegetative produce can be estimated using the following generalised equation (9) (see Table 4.5 for explanation of equation parameters):

$$FI_{veg} = \{C_{food} \times IR \times FI \times EF \times ED\} / \{BW \times AT\} \tag{9}$$

The aggregated incremental lifetime risk for exposure to vegetable produce estimated at 3.1×10^{-6} represents an annual individual risk of 5×10^{-8}. This estimate conservatively assumed the daily consumption of 80 g home-grown produce contaminated at the maximum annual concentration over a lifetime exposure of 70 years. Realistic worst-case estimates for risk contributions from this pathway, taking into account actual exposure patterns and human behavioural characteristics, were expected to be considerably lower than this upper bound estimate. Hence with reference to the HSE risk acceptability criterion, potential carcinogenic risk contributions from this pathway were regarded as negligible.

The study showed that the cumulative risk from individually investigated pathways of an individual death from carcinogenic emissions was approximately 1 in 50,000,000 and hence considerably less than the acceptable level recommended by the RCEP and HSE of 1 in 1,000,000.

The application of risk assessment, while a welcomed development to environmental assessment, should be viewed with some caution. To date assessments have not addressed all carcinogenic emissions due to the limited availability of emission data. Equally the duration of exposure and the potential unit intake of a chemical are open to conjecture. Nevertheless, the adoption of worst case assessment criteria will ensure that the impact of emissions will be significantly within currently accepted and recommended health criteria.

Questions for thought

1. What is the value of incorporating environmental risk assessment into the EA process?
2. What are the key elements of applying an environmental risk assessment?

_____

Consultation and participation

The public role in environmental assessment

◆ Introduction   90
◆ Public participation in EA in Europe and the UK   90
◆ Formal and informal opportunities for public participation in the EA process in the UK   91
◆ Strengths and weaknesses of formal and informal public participation in the UK   95
◆ Real and perceived barriers to public participation in the UK   97
◆ Future trends and mechanisms for the promotion of public participation in EA   99
◆ Questions for thought   107

Introduction

Compared with the roles of developer and environmental control authority, the role of the public in EA is not as clearly defined. EA systems and guidelines around the world vary considerably in the importance they place on the need to make information available to the public and the extent to which public groups and individuals affected by a development are formally allowed to participate in the EA process. Effective public participation demands a commitment to a two-way communication process of informing, consulting and involving the public throughout the EA process, rather than a one-way, and sometimes limited, flow of information from the developer. This is becoming an increasingly important issue as public awareness and concerns over the environment and development increase, stimulated by heightened media coverage of major developments, environmental disasters and campaigns led by the public or NGOs, which have influenced development decisions. Revealing the general design, location and impacts of the project after the ES has been prepared, or worse, after decisions have been made, is an unacceptable, and all too common, practice which frequently encourages confrontation between developer and public and restricts the clear definition of impacts and alternatives, the evaluation of significance and identification of opportunities for mitigation. Effective public participation, then, is about public and developer co-operating from the early development planning stage, prior to project announcement, land acquisition or application submission so that both acquire a clearer understanding of each other's intentions and concerns. This can lead to a development proposal which is more acceptable to all parties concerned and reduced costs in both the EA studies and construction/operation of the project.

Public participation in EA in Europe and the UK

Article 6 of Directive 85/337/EEC 'The assessment of certain public and private projects' makes provisions for public participation in the EA process, by stating that information gathered and presented in an ES should be made available to the public, and that the public concerned should be given the opportunity to express an opinion before the project is initiated. Similarly, Article 9 states that once the decision on a project has been made, the public concerned must be informed of the decision, any conditions attached to it, and the reasons for the decision. In general, Member States of the EU need to determine:

◆   who the 'public concerned' are;
◆   the places where the information can be consulted;

◆ the way in which the public may be informed, e.g. bill posting, publicity in local newspapers, or organising public exhibitions of the project;

◆ the manner in which the public is to be consulted, e.g. by written submissions, or by public inquiry;

◆ the time limits appropriate for the various stages of the EA procedure, in order to ensure that a decision is taken within a reasonable period.

The detailed arrangements in the Member States all refer to informing the public of the project, the availability of the ES and mechanisms for receiving their comments on this. However, because the EC Directive is implemented by domestic legislation in each Member State, there are considerable differences between countries in the arrangements for public participation, depending on the particular characteristics of the projects or site concerned (Stephenson *et al.* 1995). Furthermore, the actual procedures adopted are generally left to the developer's discretion. In France for example, the degree and nature of public consultation and participation is dependent on the scale and sensitivity of the project, its location, and the environmental awareness and sensitivity of those involved (Commission of the European Communities 1993). Unless a public inquiry is required as part of the consent procedure, the ES need only be published after the decision on a project is made, making the opportunities for public participation very limited. The advantages of involving the public at an earlier stage have been recognised in the Netherlands, where the process of public participation is initiated in the scoping stage and continued throughout the whole EA process.

Formal and informal opportunities for public participation in the EA process in the UK

The UK has well-established procedures for public consultation for projects subject to planning control as a result of the town and country planning system, set up by Act of Parliament in 1947. Public participation in the preparation of structure plans and local plans became a statutory requirement with the 1968 Town and Country Planning Act. Local planning authorities (LPAs) are responsible for granting or refusing planning permission for a development, taking into account representations from the developer, statutory consultees, and the general public. If the project is subject to an EA, the LPA will take into account the information given in the ES.

The formal requirements for consultation are detailed in the UK Regulations, which set out the statutory consultees who must be contacted over a proposed development (Department of the Environment UK 1989). These bodies are required to provide the developer on request with any information in their possession, which is likely to be relevant to the preparation of the ES, although they are not obliged to undertake any research to provide additional information. However, the developer is not obliged to consult any organisations at all, before the submission of an ES, and if a statutory consultee is approached for information, he or she is not required to comment formally on the project at this stage.

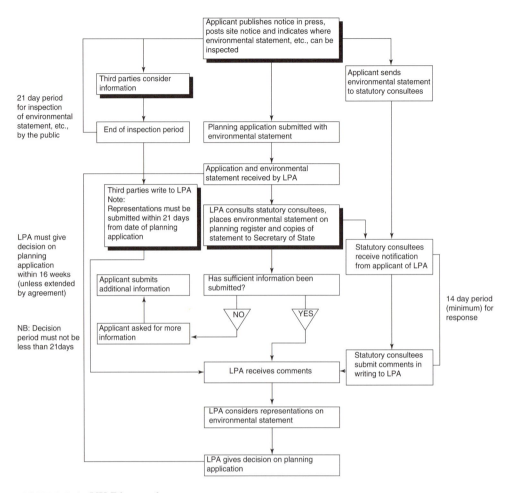

FIGURE 5.1 UK EA procedures
Source: Crown copyright. Reproduced with permission of HMSO.

The formal provisions for consultations as set out in the UK Regulations, therefore, all refer to the procedures once an ES has been submitted with a planning application, that is once the EA has been carried out. This is illustrated in Figure 5.1. On receipt of the ES and planning application, the LPA will contact the statutory consultees, who will also have received a copy of the ES from the developer. The statutory consultees then have at least fourteen days to comment on the ES in writing to the LPA. The LPA will place the ES on the planning register and send a copy to the Secretary of State. It is also at this stage that public consultation is formally required. The developer must publicise details of the planning application, and advise the public on where a copy of the ES can be inspected or obtained, through bill-posting and publishing a notice in the local newspaper. A period of twenty-one days is allowed for written representations to be made to the LPA. The LPA receives comments from the public and the statutory consultees, and must consider all representations before reaching a decision on whether to grant the project planning permission. A decision must be made within sixteen weeks of receiving the application (unless extended by mutual agreement), and the outcome of the application must be published, along with details of how the decision was reached.

The formal, or statutory, requirements for public consultation in EA in the planning system are therefore fairly limited. At no stage is the developer required to enter into discussion with the public, although the developer must ensure that access to information on the development and the ES is made available. The developer must also ensure that a non-technical summary of the ES is provided. The LPA is required to consult the public and the statutory consultees after the EA, but in the form of written submissions. There is no formal requirement for discussions or consultation during the preparation of the ES, and the emphasis is mainly on ensuring that the public is kept informed.

There is no statutory requirement for public consultation either before or during the EA process. Developers are required to publish notices of the proposed project, and make known where copies of the ES may be inspected or obtained. The competent authority is required to consider representations from the public and statutory consultees before reaching a decision, and must publicise the outcome of that decision. It is interesting that often the authority which is responsible for deciding whether an ES is required, considering any public representations, and deciding whether a project should go ahead, is also the body which is likely to be promoting the development. Examples would be the Forestry Authority giving grant assistance for planting projects, or the Crown Estate Commissioners granting a lease for marine salmon farming.

There are no statutory requirements for developers to include public participation in EA during project planning. Therefore, while government guidelines encourage participation in the preparation of an ES, a developer is under no obligation to follow them. The guidelines state that: 'The preparation of the [environmental] statement should be a collaborative exercise involving discussions with the local planning authority, statutory consultees, and possibly other bodies as well' (Department of the Environment UK 1989). The guidelines continue with: 'One of the main emphases of the process of environmental assessment is on the need for full and early consultation by developers with bodies which have an interest in the likely environmental effects of the development proposal. If important issues are not considered at a very early stage they may well emerge when a project design is well advanced and necessitate rethinking and delay. Ideally environmental assessment should start at the stage of site selection and (where relevant) process selection, so that the environmental merits of practicable alternatives can be properly considered' (Department of the Environment UK 1989).

Thus, informal opportunities for public participation are provided for and encouraged during the UK EA process, and the developer is encouraged to include EA early in project planning, for example from the project conception stage. Control of the process during these formative stages remains firmly in the grasp of the developer. Thus, if there is a wish by the public to engage more actively in the EA process, the ultimate decision is in the hands of the developer. Otherwise, the formal arrangements restrict public participation to a post-ES role. Irrespective of whether or not a development is subject to planning control, it is the developer who will decide who and at what stage and by what mechanism the public are to be involved in the informal EA process.

The notion that where the potential for EA exists then so does the opportunity for public participation, has the advantage that the development could be publicly acceptable with significant effects essentially designed out before it reaches the authorisation stage, thus possibly reducing the costs of the EA. Restricting the public participation process to the project authorisation stage in the project cycle reduces the decision-making utility, and arguably the overall value of the EA process. The next section describes some specific opportunities for public participation during the EA process.

Public participation in screening and scoping

Where the views of the LPA are sought by the developer in order to judge whether a proposal would require an ES, it has become common practice for local authorities to consult statutory consultees in screening decisions. However, the Town and Country Planning (Assessment of Environmental Effects) Regulations 1988 SI 1199 excludes public opposition as a material consideration in determining whether an ES is required, unless it is a planning matter. Despite this, there is evidence to suggest that the public do indirectly affect such screening decisions and that their views are taken into account. A study into the factors affecting screening decisions, and in particular the issue of significance, for urban development projects in England and Wales revealed that in one-third of cases the LPA was influenced by the presence of local opposition in screening decisions (Benington 1993).

Public opinion is taken into account in screening decisions by authorities responsible for authorisation of land drainage and improvement schemes and if a decision is made not to undertake an EA, this has to be publicised.

While the general public do not have a formal role in scoping decisions (the local authority should act on its behalf), public bodies with statutory environmental responsibilities are obliged to provide the developer with any information in their possession (Department of the Environment UK 1989). There is no formal requirement for these statutory consultees to comment on the proposal at this stage, although it has become the practice for some local authorities to enter informally into consultations with both developer and statutory consultees so that they may issue some guidelines or terms of reference for the developer (Wood and Jones 1991).

Public participation in the preparation and submission of an ES

Public participation in the preparation of ESs tends not to occur in the majority of cases as many developers attempt only to meet the minimum (i.e. formal) requirements. However, interest groups may take the initiative to commission and/or undertake their own EA for a proposed project, which is independent of the developers' studies (see case studies). On submission to a local authority it will be treated formally as an objection along with any other objection letters.

The submission of an ES to the local authority in the UK provides the first formal opportunity for public participation. The public are informed formally of the presence of an ES as it is placed on the planning register, advertised in the local newspaper, and they have an opportunity to inspect it and make written representations. This does not, however, extend to demanding more information from the developer if the ES is of poor quality or is incomplete. Under such circumstances the local authority acts on the public's behalf, taking into account the views of the statutory consultees. During their review of ESs, LPAs can use the comments of the statutory consultees and in-house expertise, or request supplementary information to help them reach a decision on the adequacy of the ES.

Strengths and weaknesses of formal and informal public participation in the UK

The strengths or weaknesses of a particular approach to public participation have to be gauged by its ability to satisfy the interests of the various players in the EA process. Formal opportunities for public participation have an 'inform and consult' approach and occur after the preparation of the ES and the detailed design and siting stage of the development. They therefore prevent the inclusion of different interests into the design of a project at an early stage and reduce the possibility of this occurring at any later stage since the project will have gained considerable momentum upon the preparation of the ES.

Where projects are subject to planning control the public and interest groups are informed of the development and the ES in three ways: the planning register; site notices and bill posting; and advertisements in the local newspaper. The statutory consultees are informed by the same process, although they may be notified earlier if the developer approaches them for information for inclusion in the ES. Such an approach keeps down costs and the time required by the developer. The newspaper advertisement has the potential to reach a wide audience and site notices inform the public within the immediate vicinity of the development. It is uncommon for members of the general public routinely to examine planning registers held within the LPA planning offices, although these do tend to be checked by environmental pressure groups as a means of monitoring new developments and likely areas for concern. The LPA is passive in this process and therefore minimises costs and time.

The statutory consultees are given fourteen days to comment on the ES and the public and interest groups twenty-one days. Although timescales are fairly restrictive, extensions are permitted provided there is mutual agreement in order to accommodate the statutory consultees who may feel unable to comment on the quality and substantive nature of the ES in such a short timescale. A short timescale may also restrict the number of people able to consult the ES, as several hours may be required to read beyond the non-technical summary. Against this, short timescales are beneficial to the developer who incurs costs in any delays.

The formal opportunities allow for the LPA to make informed decisions as it receives information from the developer, the general public, the interest groups and the statutory consultees. As the LPA is required to publish the rationale for a decision, the statutory bodies and the general public will be able to see how their concerns have been taken into account in the decision-making process.

The only information available to the general public, the statutory consultees and the interest groups is the ES and the non-technical summary. In some cases, this may be insufficient to serve the interests of the different players, especially the general public who may face difficulties in dealing with a complex technical report and who have no formal scope for dialogue with the developer. ESs are usually held at the LPA offices and local institutions such as public libraries. The public must seek out the ES in order to read it. This means a fairly limited access to the documentation and it is a general observation that very few people actually consult an ES written for the project they are concerned with. The only formal way of expressing a concern is to make written comments to the LPA. Members of the public are not permitted to enter into discussions when the LPA Planning Committee makes its decision, and hence it is difficult for them to contribute to problem solving or making modifications.

The uptake of informal opportunities for public participation is developer dependent, discretionary and may be designed to satisfy the developer's needs more than the interests of other players. Two factors need to be considered in the design of public participation mechanisms: the exercising of the informal opportunities; and the approach adopted (that is, inform, consult or involve the public). Exercising the informal opportunities which fall early on in the project cycle helps to serve the interests of the various players at a stage when alterations to a project design to accommodate environmental and socio-economic concerns are more feasible. Mounting a public participation exercise in addition to those required by the formal process may be more costly, and this cost usually has to be borne by the developer. Adopting an approach which involves rather than simply informing is also going to be costly and more time-consuming for all concerned.

The developers in the case studies described in this chapter used different approaches to public participation. Tarmac plc used public display and public presentations at the detailed design stage and allowed comment on the ES prior to the submission to the local planning authority. Southern Water Services used public displays and presentations at the scoping stage of the ES to involve the public. North West Water used discussion and presentation and allowed the local authority and the statutory consultees to influence the decision on the choice of the alternative.

Leaflets and exhibitions are a good way of reaching the public if there are sufficient leaflets to inform a large number of people and the exhibitions are accessible to all those concerned. Given that the information provided in exhibitions and presentations is very much in the control of the developer, there may be a feeling among members of the public that it represents little more than a public relations exercise for the development. Such mechanisms for informing the public can be carried out at a relatively low cost and are less time consuming than interactive meetings. However, they are effective only if the public receive the information they need to satisfy their interests. Providing information by way of exhibitions and leaflets is a one-way flow of information and does not allow the public to discuss their interests and understand the issues involved. For this to become a two-way communication process, follow-up meetings or more information may be necessary.

Mechanisms used by the developers or their consultants in the case studies described included meetings, questionnaires and written responses to consultation documents. Such mechanisms are fairly time-consuming, but allow for discussion of the proposals and therefore a greater degree of understanding. Unless it is made clear how information gained from the consultation is going to be used in the decision-making process, the public are likely to remain sceptical of the proceedings and may still oppose a development on those grounds. Questionnaires are useful instruments to orientate demographic and social variables, but not to discover in-depth views or reasons for these views.

Once the consultations have been completed, the issue is how to incorporate all the comments into a form which is suitable to aid decision making. Analysing a large number of responses is time-consuming and presents the difficulty of the analysis remaining objective, particularly where a summary of wide-ranging comments is desired. The consultants in the case study of gold mining in Ireland transcribed all the comments made at the meetings from tape recordings and then used statistical analysis in an effort to remain objective.

The developer still retains control over the consultation mechanism and may have great difficulty in collating often contradictory responses. The mechanism adopted may preclude dialogue between different interest groups with an emphasis on communication between the developer and the other groups.

Real and perceived barriers to public participation in the UK

The current provisions of the EC Directive and the UK regulations are undoubtedly a major barrier to public participation given that the law limits this to the post-ES stage of the EA process. The procedures for informal public participation are largely controlled by the developer and there is no officially recognised review body in the UK to oversee the process and provide advice to all those concerned.

Cost and time are limiting factors to the choice of mechanisms for public participation and as this cost is incurred by the developer, this is a barrier to exercising discretion of informal opportunities. The EA practitioners or consultants who research on behalf of the developer have to remain competitive if they are going to survive in an ever more constrained market place and therefore may be unwilling to specify the work they would do which is beyond that of the legislation, for risk of losing the contract in the tendering process. Statutory consultees also have time and financial constraints which affect their ability to respond effectively to the ES. The general public and interest groups would have difficulty in participating in mechanisms requiring extensive discussions because of other commitments and a lack of resources. Perhaps most importantly, developers do not tend to see the value of informal public participation as a means of cost saving and/or quality improvement in project design. If such value were more directly apparent, for example by more rapid planning approval, developers (and their consultants) might engage in public participation activities more readily.

The EC Directive and the UK regulations provide no guidance on how to carry out public participation. Wood and Jones (1991) found that there was a widespread lack of familiarity with EA procedures on the part of all players in the process and it is doubtful that many of the general public even know what EA is. Public participation in the UK has traditionally been based on a decide–announce–defend model and there is more familiarity with adversarial approaches such as public inquiry and appeal mechanisms. Interest groups can be very adept at using the system to achieve their interests and remaining an equal power in it.

The ability of the general public and interest groups to participate in board room type discussions is also a barrier that is dependent on the skills of the public to articulate their interests. Such an approach tends to exclude those people who are socially disadvantaged depending on their race, education, religion or gender and are traditionally less able to attend such exercises because of social barriers.

As a public document, an ES should be written in a style which enables readers with different interests to understand for themselves how its conclusions have been reached and to form their own judgements on the significance of the environmental issues associated with the project (see Chapter 6). This may be an ideal which is difficult to meet given the technical complexity of many projects and their environmental effects and may present itself as a barrier to effective public participation.

Because of the desire to protect the commercial confidentiality of a project, a developer may be reluctant to provide public access to information entrusted to the LPA until the law requires it.

Developers often regard EA as an unwelcome obstacle in the way of obtaining planning consent and are wary that, simply by their involvement in the process, they tend to imply that their development is environmentally suspect. Developers and interest groups commonly have a general mistrust of each other. The public and interest groups may believe the developer to be economical with the truth in its public relations exercises and the developer may be reluctant to engage in public participation in case the public raise unreasonable demands, or are politically motivated.

Such views obscure the potential of EA to improve the project design in a manner that can benefit the developer. To further ignore the information available from the public is to assume that the developer has access to all relevant knowledge at the outset.

The LPA, statutory consultees, developer and consultants may all believe that they have the professional competency to act fairly on behalf of the public in the EA process. They may see public participation as an unwelcome intrusion into their area of competence. Some planners may regard informal public participation as lobbying and any compromises or compensations gained by the community without the involvement of the local authority as an intrusion into the procedures for planning gain.

Lack of resources is a major barrier for public participation in the EA process, particularly for interest groups forced into an adversarial role during the public inquiry system. At present, there are no arrangements for funding of interest groups who are preparing a case at public inquiries. The Council for the Protection of Rural England (CPRE 1991) believes that financial support for interest groups is justified because of the influence public inquiries have on government policy.

Future trends and mechanisms for the promotion of public participation in EA

Public participation, sustainable development and the planning process in the UK

Agenda 21 (UNEP 1992) stressed the need to empower people and promote a people-led approach to sustainable development. The UK's Strategy on Sustainable Development (Department of the Environment UK 1994) details how the UK will, over the next twenty years, work towards the goal of sustainable development. This highlights the Town and Country Planning System as a key instrument in delivering land-use and development objectives that are compatible with the aims of sustainable development. Environmental appraisal techniques are seen by the government as helping the planning system deliver these objectives and the government acknowledges the need for more public participation in decision making, in response to comments made to it:

> The role of voluntary organisations and the general public were subject of a number of responses, they ranged over the need for better funding for non-profit making organisations to take a wider role in raising awareness of environmental issues; more structures for informal and formal participation in decision making; and suggestions for ways in which local communities and groups can get involved in voluntary initiatives. Many pointed out that the voluntary section was uniquely placed to take sustainable development forward in broad and issue led ways by using a variety of approaches and developing cross-sectoral partnerships.
>
> (Department of the Environment UK 1994)

Clearly the UK Strategy will have consequences for the decision-making process in the assessment of new development and will create a trend which moves towards involving the public in decision making and developing mechanisms to build consensus between the players in the EA process. One indication of this is seen in the remit of Scottish National Heritage (SNH) to ensure that anything affecting nature conservation, recreation and landscape in Scotland is sustainable. Since SNH is a consultee in EA it is already required by law to advise the local authority on sustainability issues as they arise in EA.

Some mechanisms used to promote public participation in EA

Community right to know is an informing mechanism and, by itself, provides the weakest form of public participation. It does, however, help to remove barriers caused by a lack of information, and encourages a more open EA process. There is no opportunity for dialogue, but it does provide members of the public with the information they need to make up their own minds. The exercising of the right to know would therefore reduce the control the developer has over the amount of public participation in the EA process, since less would be left to the discretion of the developer.

Historically, secrecy has been endemic in environmental legislation, as many statutes have contained specific sections explicitly forbidding disclosure of information relating to environmental issues (Ball and Bell 1991). The main reason for this has been that free public access to environmental information could affect the viability of industrial operations. The UK has registers of environmental information held by public bodies which must make that information available to anyone on request. The main driving force for this turn-around in government policy has been the need to comply with EC Directive 90/313 on the Freedom of Access to Information to the Environment. Such registers and greater freedom of access to information may reduce public opposition to projects where this has been fuelled by mistrust of the developer or lack of information. Because such registers are in place only for operational projects, such policy trends will not improve public access to information in the UK formal EA process although they may aid public participation in the informal process, especially in the auditing and monitoring of new development.

Opportunity for third-party appeals and reviews of decisions

Local planning authorities (LPAs) in Great Britain have the right to determine planning applications for new development. In the event that an application is refused or the LPA fails to determine the application in a specified period of time or the application is permitted subject to conditions opposed by an applicant, the applicant has the right of appeal to the Secretary of State. In exceptional circumstances, where for example the proposed development is of national importance, a planning application may be 'called in' for determination by the Secretary of State.

Third parties have no right of appeal in the UK. They must apply for judicial review of any adverse decision; this is costly and they are unlikely to win (Purdue 1991). The right of appeal rests with the developer. The onus in the UK is on the authority to justify refusal of planning permission and the only role the public have is in determining if this is upheld.

Opportunities for public inquiries

Before making his decision, the Secretary of State will require the applicant, the LPA and interested parties to submit their respective cases to him in a written form or to present their case at a public inquiry. An Inspector (Reporter in Scotland) who hears evidence from all sides including the developer, the LPA and interested groups, chairs the inquiry. Each group is cross-examined on its evidence and its professional competency to give such evidence. In all but major forms of development, the Secretary of State delegates the decision to the Inspector.

Use of alternative dispute resolution

Alternative dispute resolution, a mechanism designed to achieve agreement and consensus between the parties involved, is not commonly used in the UK, although there has been some use of environmental dispute resolution, a process whereby interested groups are invited to participate in negotiations before any final siting decision is made (Purnell 1992). Such an approach allows a mutual understanding of the different values that underlie the conflicts and the development of alternatives that serve this range of interests, and allows for discussion over factual disagreements. The success of a meeting geared to this approach depends on the way the members feel they have been treated in the proceedings and the equity which has been achieved in the sharing of power. Such mechanisms, however, have methodological problems and cost and time constraints and require a mediator or facilitator to conduct the proceedings.

A technique known as Requisite Decision Modelling was used to help identify a sustainable land-use plan on an estate in the Scottish Uplands (Moss 1994). The technique allowed the different interests and interpretations of sustainable development of the members of the research group to be computer modelled. This predicted different outcomes depending on the ranking of those interests, and so allowed them to predict a range of different futures and decide whether those were sustainable.

The main advantage was that it is a quantitative and transparent system capable of integrating all types of available information into decisions about land use and is an impartial mechanism for generating consensus for divergent opinions (Moss 1994). The main disadvantage of such an approach is that it requires careful organisation and is very time consuming. A mechanism like this has utility for public participation in EA where sustainable development is a goal and is applicable to the conflict resolution of complex problems.

CASE STUDY: GOLD MINING IN CONNEMARA AND SOUTH MAYO, IRELAND

During 1989, a number of exploration and development companies were proposing to exploit gold reserves in West Connacht in Western Ireland. These proposals caused concern among local communities as well as regional and national organisations, resulting in the formation of a committee called Gold-EIA, whose objective was to address the lack of information available to the public on the environmental, economic and social effects of the development and to provide an objective base for its members and other interested parties. Gold-EIA commissioned CEMP, based in Scotland, to undertake a scoping study which would identify the key issues and form the terms of reference for a subsequent, comprehensive EA study. Central to this study was an analysis of the views and concerns of local residents and other interested parties (CEMP 1989).

Following an identification of interest groups, a consultation document inviting written comments was prepared and two public meetings organised for the expression of views and concerns. The consultation document gave details of project activities and likely topics to be included in the EA. It invited written responses to a list of topics and on any other issues which the individual or public body believed to be either important or irrelevant. An explanation was given as to how comments from respondents supplying names and addresses would be used.

At the meetings, individuals were able to make a verbal representation of their concerns. These were taped, transcribed and compiled as a number of common issue categories, whose frequency of occurrence was recorded as a percentage of the total number of issues. In this way, the relative importance of the issues could be demonstrated and used to form the basis for the terms of reference of the EA. The key issues generated by this study were tourism, fisheries, mariculture, employment, religious significance, quality of life and local livelihoods. The Irish government opposed prospecting for gold, and won a case in the High Court to implement a presumption against mineral development. Subsequently, Gold-EIA did not progress to the production of its own EA.

CASE STUDY: SAND AND GRAVEL EXTRACTION, BEDFORDSHIRE, UK (STEPHENSON *et al.* 1995)

In 1990, planning consent was sought by Tarmac for the extraction of sand and gravel at a site north of the village of Broom, Bedfordshire, UK. The development was anticipated to last for a period of between nine and thirteen years depending upon an extraction rate of up to 500,000 tonnes per annum, followed by restoration of the site to include fishing and sailing lakes, footpaths and bridleways, and provisions for nature conservation. The developer held a public display/exhibition and made presentations at local public meetings prior to making a formal planning application and completion of the ES. Local residents in Broom and in neighbouring Upper Caldecote voiced an overwhelming opposition to the development and established an action group, Bedfordshire Campaign Against Gravel Extraction (CAGE), which then carried out an EA independently of the developer and submitted this to the planning authority, Bedfordshire County Council, as part of its objection to the development. Having taken into account the ES and other representations prepared by CAGE alongside the application and ES from Tarmac, the planning authority granted planning permission in 1992. Although CAGE was unsuccessful in preventing the development from taking place, its activities led to a modified version of the original application which included a number of planning conditions, including a reduction in time for extraction from ten years to nine years, improvements in screening between

Broom and the extraction site and a doubling of the distance between village and site. Changes in phasing of extraction were also required to ameliorate disturbance to the village. Changes in the post-extraction restoration plan were required to improve access around the site and the provision of a car park for the fishing lake.

CASE STUDY: THE FYLDE FORUM

A proposal to construct a sewage treatment plant and long sea outfall to discharge treated sewage at Fleetwood, Lancashire, on the north-west coast of England was made by North West Water (NWW) in 1991 (Stephenson *et al.* 1995 and North West Water 1992). This was a substantially modified version of an original scheme proposed in 1988 which was rejected by the UK Department of the Environment following a public inquiry. The grounds for refusal were that discharging screened but untreated sewage into Morecambe Bay would not meet the requirements of the forthcoming EC Directive on urban wastewater treatment, which called for full primary and secondary treatment of sewage for new sea outfall schemes serving a population of over 150,000. Also, there was the question of the initial proposal not meeting the EC Directive on Bathing Waters.

The development would intercept sewage from the existing five short outfall locations and pump it to a treatment plant at Fleetwood (primary and secondary treatment). The plant would have large capacity stormwater storage tanks, permitting virtually all storm flow to be treated. Final discharge of treated sewage would be by a long outfall into the Lune Deep.

The scheme originally proposed in 1988 aroused widespread public opposition, although it did receive planning approval from the local planning authority before being turned down by the UK Department of the Environment. This public opposition arose at least in part from a failure of NWW to seek out public involvement in the development's design and site selection procedure.

As part of the planning for the revised 1991 scheme, NWW drastically changed its ways of informing the public and established a public working party to help determine a more acceptable proposal prior to seeking planning permission. An advisory committee called the Fylde Forum was set up by NWW, including representatives of the local community; politicians; planning officers from the district and county councils; local industrialists; representatives of environmental pressure groups; and representatives of the National Rivers Authority (NRA). Members were included from the whole area, from Preston in the south to Morecambe Bay in the north. Forum members themselves decided how it should be run and arranged to hold meetings every two months in which a number of alternative schemes and opposition to these were discussed. A sub-committee was established (Technical Officers Working Group) in order to interface with NWW with respect to the technical details of the scheme. In-house and external specialists hired by NWW drew up a total of fifteen alternative schemes and presented a brief on each to the Forum, which then proceeded with debating and determining a preferred option. Although much of the work was done by the Technical Officers Working Group, the full Forum took all the decisions.

The original intention was that a single preferred option would be chosen. However, the often conflicting concerns of individual members and the fact that the Forum was not elected to represent the public's views led instead to the selection of a short list of three alternatives drawn up after four meetings and presented to NWW. These were: to take all sewage from existing short outfalls for treatment at Preston some 10 km in from the mouth of the Ribble Estuary; to take all sewage for treatment and discharge from Fleetwood; and a combined Fleetwood/Preston option.

Although the final decision on the location and design of the sewage treatment and disposal scheme was taken by NWW, this was considered to be the most appropriate action to be taken by the Forum itself. The case study serves as a rare example of a self-running scoping committee, consisting of representatives of various local interests (albeit unelected) whose activities were fully supported by the proponent. The early participation of the public and other interested parties, prior to any submitted ES and planning application, led to an early alleviation of fears and concerns over the environmental impacts of the development and a reduced likelihood of formal objections being made.

CASE STUDY: TESTWOOD LAKES

In 1991, planning consent was sought by the Hampshire Division of Southern Water Services Ltd for a ten-year project in which sand and gravel would be extracted from a 58 hectare (ha) site, north-west of Southampton, Hampshire, followed by the phased construction of storage reservoirs (Southern Water Services Ltd 1991). The developer made the case that the reservoirs were needed for emergency supplies and to provide general storage to meet a forecast increased water demand in southern Hampshire until 2015. Prior extraction of minerals was necessary to obtain the maximum volume of water storage, and to avoid sterilising a valuable resource. The development was expected to have significant impacts, involving land-use change, visual impacts, noise during the extraction phase, and changes to the hydrology and ecology of the area, especially affecting the River Test. However, it was also expected to provide positive opportunities for recreation, local amenity, and nature conservation.

Consultations were carried out on behalf of the developer with official bodies, interest groups, local residents and the general public to assist with the scoping for the EA. All these consultees were offered the opportunity to comment before designs were finalised. The possibility of alternatives to the scheme, such as water conservation initiatives, desalination works, or using existing lakes and ponds, as well as alternative sites for the project were discussed with specialist consultees, but not with the public. The consultations, using a series of presentations, informal meetings and a mobile public exhibition of the draft proposal, provided views from all concerned on the nature of the development, the site to be used and the main impacts (considered to be landscape, traffic, noise, ecology, hydrology and water quality) and these were addressed in the final ES.

Few objections to the project were raised and planning permission was granted without the case going to appeal. The general acceptance of the project is indicative of good consultation procedures, in which the concerns of affected parties were sought early on in the EA process, and taken into account in the project design to make it more acceptable.

CASE STUDY: SOUTH WARWICKSHIRE PROSPECT

In December 1985, British Coal proposed to develop a deep coal mine in South Warwickshire. The developer commissioned consultants to undertake an EA which included the selection of a mine site within the prospect area together with a range of spoil disposal options (Environmental Resources Ltd 1987). As part of the EA process, a forum was held involving officials from British Coal and representatives of five local authorities which sought to exchange information, reach agreement of fact, and identify a site and appropriate infrastructure. These discussions led to the selection of eight potential mine sites. Other public consultations were undertaken through an extensive series of presentations at meetings attended by representatives of interest groups and statutory consultees in order to assist in the selection of a preferred mine site. British Coal produced a leaflet entitled 'South Warwickshire Prospect' which was delivered to residents of local properties in January 1986 and produced a number of press releases about the scheme. The consultants carried out their own consultation exercise which consisted of a short consultation letter being circulated to interested parties and the circulation of leaflets to the local community which invited feedback on any environmental issues they wished to raise. The consultants also held meetings for those wishing to express their views.

The results of these meetings, the written responses and the British Coal consultations were used to help identify a set of criteria which reduced the number of alternative sites from eight to five. An EA was undertaken on these five sites and two options selected for the further assessment of technical and economic constraints. The preferred option selected was at Hawkhurst Moor. Subsequently, the proposal went to a public enquiry where it was turned down.

CASE STUDY: WASTE MANAGEMENT IN ALBERTA, CANADA

One area that clearly shows the public role in EAs, and for which there are many examples of success and failure, world-wide, is the siting and operation of waste management facilities. These frequently create public reaction and provide a good basis for developing a rationale for public involvement. This case study (McQuaid-Cook 1994) concerns the siting of a hazardous waste treatment facility in Alberta, Canada, and demonstrates how a public participation programme can work within the EA process to find a more acceptable solution to the problem of waste management. The ideal situation of no preconceived idea of the best site for the facility meant that various sites could be compared, and the community was not restricted to one area. Prior to any EA, all regions of the Province of Alberta were visited by a government study team and the issue of waste management was discussed at meetings with the local government and the public. As absolutely no decisions had been made on siting, public response was excellent, and rational, rather than emotional, conversations were held with all interested parties.

The linking of siting, EA and the public participation created a strong connection between science and culture. All information developed during planning and implementation phases was made available to the public, and liaison was on-going with all interested communities. Results of studies (such as soil testing, water analysis, wind modelling, archaeological surveys) were provided to local councils and involved community groups.

It was found that the commonly used technique of communication, the large public meeting, was counter-productive. Community members who opposed the project turned out in force, became overbearing with their statements of harmful impacts and created an uncomfortable atmosphere for those who had genuinely attended to learn about the project before making a decision. Many citizens were too shy to ask questions in front of a large group. The presenters had difficulty inter-relating with the audience, so many questions went unanswered. During Alberta's programme, changes were made to the public participation process that resulted in success. Small meetings replaced the large open forum and presentations changed from lectures to discussions. This entailed much more work for the project team, but the results were tangible. By limiting the number of people in attendance at meetings, duplicate meetings had to be held until all interested people had attended. In this way, a more intimate rapport was established, questions answered, points explained and an understanding developed among all parties, whether they were for or against the project.

Local liaison committees were established to serve as contact points for both the proponent and the public. Members of the committees had full access to environmental data, project site plans and background information. Representatives from the committees were invited to visit similar facilities elsewhere and report back to their communities. They were included in seminars and conferences on waste management, and ultimately had the opportunity to be represented on the board of both the government corporation and the facility operating firm, joint partners in hazardous waste management in Alberta.

All information during the siting programme was available through fact sheets, reports, background papers, maps and press releases. Candidate communities were invited to critique all site selection materials. Local newspaper editors were invited to attend meetings for updates on progress and materials were available to them for publication in their papers. Environmental interest groups were invited to meetings to discuss their concerns. Representatives from candidate communities were invited to workshops to share their questions and ideas. Graphic materials were prepared to ensure that all parties were kept up-to-date on all aspects of the site selection process and the bio-physical, economic and cultural aspects of the EA.

As a result of the intensive public consultation and participation in Alberta, a hazardous waste management facility was successfully sited, based on both environmental suitability of the site and local community approval. Waste management is one of the most difficult topics to deal with in the public forum, but once the people understand the impacts of previous waste management practices in their region, the types of waste which may be hazardous, the amounts generated in the home and local businesses, and the environmental consequences of doing nothing, they can then make a rational decision on how to proceed in their community. As a result of the strong public participation, Alberta now has a satisfactory system of waste handling facilities throughout the province which effectively deals with recycling, treatment and disposal.

Questions for thought

1. What is the value of public participation in the EA process?
2. Identify ways in which the public can be consulted in the EA process.
3. What is the difference between formal and informal public participation?
4. How can the media (television, radio and newspapers) be used to promote public participation in the EA process?

Chapter 6

Managing the
EA process

◆ Introduction   110
◆ Context and procedure   111
◆ Technical management   111
◆ Report writing   116
◆ Financial control   119
◆ Questions for thought   125

Introduction

While it is recognised that EA, when properly applied from the earliest stage in the planning process, can accelerate development (Mudge 1994), ineffective EA project management can delay the process. Managing the EA process is fundamental to the entire satisfactory delivery of the ES. Failure to conform to a systematic procedure may lead to a breakdown in the communication process. The prerequisite of any EA process, EA management involves information management and its communication, and unfortunately there are many opportunities for mis-communication. Bingham (1992) has identified three fundamental aspects of EA and project management: context and procedure, technical management, and fiscal control and budgets. To this can be added the importance of report writing (Canter 1996).

EA is not a one-off process ending with the production of an ES. It should provide an essential input to project management through a continuing evaluation and re-evaluation of the various environmental issues, as project plans are developed, defined and refined. This process should continue throughout the life of the development from conception to final abandonment or closure and reclamation of the site. It is important, therefore, that careful consideration is given to the scope, management, planning and financial aspects of the EA process.

In the absence of clear direction from the project proponent and/or authorising body, an EA may not cover all the relevant environmental issues while pursuing other less important matters. In such circumstances, requirements of the various interested parties may not be met. At the outset, therefore, terms of reference (TOR) should be prepared defining the key issues to be covered, the decisions to be taken and the options to be investigated so as to minimise omissions, and the possibility of the introduction of costly additional issues in the latter stages of the EA process. This is particularly important where an EA is required in response to legislative requirements. It is critical to recognise that the technical validity of EA studies can easily be compromised without effective study management.

The goal is to achieve management of the environment through the EA process. If a project manager is to achieve this goal, it is necessary to look beyond the product he or she is in charge of managing. Reports sit on shelves and collect dust. Their utility is only as good as the commitment to action, perhaps inaction, that they generate. One can sum up the project management of EA in one word – COMMUNICATION.

An EA study may take a long time to produce, and therefore the project manager will be responsible for the continuity of the project despite possible project team changes, changes in the tender brief, etc. According to the World Bank (1991) EAs take as much time as a feasibility study; in other words they can take from less than 6 months to more than 18 months to complete. The longer the duration of the project, the greater the potential for managerial problems.

Context and procedure

An initial phase of the EA management process is the understanding of the manager's role in the context of the project and environmental procedures. In some situations the context and procedures are well defined and are established in regulations, law or policy (see Chapter 1). In such circumstances the project manager's role is often well, or partially, defined. However, in countries where EA procedures are less well defined there is a greater opportunity for flexibility and creativity for the management process. Indeed, EA procedures can in themselves be ignored. Bingham (1992) identifies five crucial steps that can benefit a project manager before commencing the management process. These are summarised in Table 6.1.

What one must not forget is that the environmental process provokes question-asking. Both the questions and the answers can lead to a different focus for a project, perhaps even clarify its purpose and need. The EA project manager serves an important role in asking questions and resolving them. The five essential project manager questions are: why, when, who, what and where? Answering each of these questions is critical to a selection of analytical procedures and methods.

If the practitioner can tackle each of these questions at the outset, the technical aspects of project management are made easier. By answering the why, when, who, what and where and sharing the results with staff, one can focus the technical work on the subjects and areas of relevance.

Technical management

For a major development, or for smaller projects of high and diverse environmental sensitivity, multi-disciplinary teams are usually required to prepare an ES. Typically, such disciplines may include economics, environmental management, agriculture, forestry, water resource management, atmospheric sciences, etc. Table 6.2 sets out some of the specialists that would be called upon according to the World Bank (1991).

For larger projects, an overall co-ordinator and, in some instances, a management or steering committee, may be appointed to:

◆ ensure that all disciplines are working to a co-ordinated brief (e.g. TOR);
◆ act as the contact point with project management and others;
◆ identify the need for specialist investigations and their inclusion;
◆ programme all activities so that information is available at the appropriate time for permit applications and meetings; and
◆ organise and collate team members' input to produce an EA document for management, authorities and others.

Choosing the team to work on an EA is desirable, but choices are often a matter of who is available, who can be afforded and how many people can be chosen or assigned within the budget. For the project manager, a frequent factor undermining confidence is cost.

TABLE 6.1 Review questions to be asked prior to undertaking the EA management process

---

Why do EA? In the absence of regulatory or other procedures that give the project manager the answer to why an EA is needed, the project manager must ask: Who decides the need for an EA and who evaluates or, perhaps, influences the need for an EA? Answers will, at the least, provide the project manager with some insight into the objectives of his management and may contribute to the rationale for requesting, questioning the need for or defining the kind of EA to be undertaken.

When does EA occur? The environmental process ideally begins early, as early as pre-feasibility. Thus a project manager may be called upon to initiate or carry out a preliminary step, sometimes referred to as an environmental sensitivity study, an environmental constraints analysis, an environmental scoping study, or an initial environmental examination. Its purpose is to identify what the major problems might be and to determine whether they can be resolved at that stage, or whether further environmental analysis is needed.

Who is responsible? As already noted, an EA project manager is among many responsible participants. The project manager should be familiar with the list of other interested parties in the EA process. Asking the question 'Who am I in the EA process?' will not resolve conflicts, but it will create an awareness of the range of interests and enable appreciation of multiple perspectives. Ascertaining what role one plays in the EA is more than an exercise in interpersonal dynamics. By seeing multiple perspectives, the project manager can better help in negotiating acceptable solutions. The project manager can also review the list of interested parties to ask what each wants and what their rules, jurisdiction and interests might be. Wherever possible, the project manager, should establish, through interviews or meetings, some rapport with the interested parties. Establishing this communication process can be accomplished at the same time as the project manager and his staff gather information about the project or the environment.

What is the action? Among the easiest mistakes the EA project manager can make is to concentrate on the characteristics of the existing environment and neglect to pursue detailed knowledge of the proposed action. Without consideration of how the action alters the environment, there is, by definition, no impact assessment. To accomplish an understanding of the action, the EA project manager must communicate with programme developers, project designers, engineers, and whoever else has formulated the action. The project manger should be able to identify the project components, alternatives and scoping and defining project issues.

Where are the zones of impact? Establishing the zone of impact is the project manager's responsibility. A common mistake that an environmental expert may make is to concentrate too much on the site of the activity itself and not on the surroundings. Again the project manager needs to make these study areas common to as many disciplines as is reasonable.

Source: Bingham (1992)

TABLE 6.2 Specialists likely to be used in EAs

Natural resource	Subcomponent	Specialist
Air	Air quality	Air quality/pollution analyst
	Wind direction	Air pollution control engineer
	Temperature	Meteorologist
	Noise	Noise expert
Land	Land capability	Agronomist
	Soil resources/structure	Soils engineer
	Mineral resources	Soils scientist
	Tectonic activity	Civil engineer
	Unique features	Geologist
		Geotechnical engineer
		Mineralogist
		Mining engineer
		Engineering geologist
		Seismologist
Water	Surface waters	Hydrologist
	Groundwater regime	Water pollution control engineer
	Hydrologic balance	Water quality/pollution analyst
	Drainage pattern/channel	
	Flooding	
	Sedimentation	Civil/sanitary engineer
		Marine biologist/engineer
		Chemist
		Hydrogeologist
Flora and fauna	Environmentally sensitive areas	Ecologist/Forester/Wildlife
	Species inventory	biologist
	Productivity	Botanist
	Biogeochemist/nutrient cycling	Zoologist
		Conservationist
Human	Social infrastructure/institutions	Social anthropologist/Sociologist
	Cultural characteristics	Archaeologist
	Physiological and psychological well-being	Architect
	Economic resources	Social planner
		Geographer
		Demographer
		Urban planner
		Transport planner
		Economist

It can therefore be seen that EA studies are often conducted by interdisciplinary teams, i.e. a group of two or more persons trained in different fields of knowledge with different concepts, methods, and data and terms, organised to address a common problem with continuous inter-communication among participants from different disciplines (Dorney and Dorney 1989). A study team for a specific environmental impact study can be considered as a temporary entity which has been assembled, and possibly specifically appointed, for meeting the identified purpose of conducting an EA for a proposed project. The team may be assembled with formal authority, responsibility, and accountability, but a more typical approach is the delineation of an informal authority within that team, with the team basically being subjected to the management of the team leader (Cleland and Kerzner 1986).

The number of members of an inter-disciplinary team can vary from as few as one or two to perhaps as many as eight or ten individuals or more, depending upon the size and complexity of the EA and the project budget. Poor team skills will undermine the success of management and completion of the ES. In selecting the study team, the project manager should take into consideration the following substantive needs and characteristics of individuals (Canter 1991b):

◆ the types of expertise needed relative to the EA;
◆ the experience of the prospective team members on similar or other types of projects;
◆ the orientation of the individual towards working with other individuals on a group effort;
◆ the receptivity of individuals to the viewpoints of other disciplines;
◆ the range of interests of the individual, with a broader range of interest being more conducive to successful work on an environmental impact study than a narrow or limited range;
◆ availability within the overall work unit time schedule to work on the team; and
◆ some indication of the following work traits and personal characteristics:

    1.   organised;
    2.   orientated to work on a time schedule;
    3.   no aversion to writing;
    4.   willingness to travel and make site visits;
    5.   willingness to work with other individuals and serve as a team player;
    6.   self-starter;
    7.   creative;
    8.   expertise related to the local geographical area;
    9.   adequate verbal and written communication skills;
   10. credibility with other professionals in the field; and
   11. adaptability.

The project manager is the individual who provides leadership for the team itself when directed towards accomplishing the end purpose, with the end purpose being the successful execution of the EA study (Cleland and Kerzner 1986). The project manager should exhibit a number of specific qualities; required attributes should include (Cleland and Kerzner 1986):

◆ demonstrated knowledge and skill in a professional field;
◆ positive attitude in support of the conduct of the environmental impact study;
◆ a rapport with individuals;
◆ an ability to communicate with both technical and non-technical persons;
◆ pride in his or her area of technical expertise;
◆ self-confidence;
◆ skill as a self-starter;
◆ a reputation as a person who gets things done;
◆ the ability to deal successfully with the challenge of doing quality work; and
◆ the willingness to assume responsibility for the overall study and team leadership.

In summary, several key characteristics should be considered in the selection of the project manager. These characteristics include, in order of priority:

◆ experience in serving as project manager;
◆ management/leadership skills; and
◆ substantive area of expertise.

A number of considerations are related to the management of a study team and an EA study. The project manager should consider several management techniques and develop approaches to utilise them for the successful operation of the specific team. For example, Cleland and Kerzner (1986) suggested the following important factors which would be basic to the successful management of a study team:

◆ a clear concise statement of the mission or purpose of the team;
◆ a summary of the goals or milestones that the team is expected to accomplish in planning and conducting the EA;
◆ a meaningful identification of the major tasks required to accomplish the team's purposes, with each task broken down by individual;
◆ a summary delineation of the strategy of the team relative to policies, programmes, procedures, plans, budgets and other resource allocation methods required in the conduct of the environmental impact study;
◆ a statement of the team's organisational design, with information included on the roles and authority and responsibility of all members of the team, including the team leader; and
◆ a clear delineation of the human and non-human resource support services available for use by the study team.

A fundamental technique for team operation is the holding of periodic team meetings with planned agendas. It is a primary role of the project manager to develop schedules and to establish priorities with regard to manpower and other resources allocated to particular activities within the EA. It should be recognised that modifications will probably be needed in scheduling and budgetary allocations as the study progresses. These are typical in EAs.

In addition to team meetings, the project manager must allow individual team members working in their own particular areas to carry out agreed assignments, and then subject the work products, or at least the ideas resulting from the work, to team review. The pattern of meeting, individual work, and a follow-up review meeting is a useful concept in the operation of a study team. While it is theoretically possible, it is unlikely that the study team will work completely together on every aspect of an EA.

One of the issues which is often related to team management is associated with the periodic necessity for having special studies conducted by experts who are not members of the study team. An example might be the undertaking of specific cultural resources surveys by archaeologists. If special studies are required, and they are common in EAs, then the team management concept should include a meeting to discuss the requirements of the special studies, the particular TOR for groups or individuals to conduct such special studies, and the clear delineation of the anticipated output from the studies, with particular care given to ensuring that the output from such special studies will coincide with the needs of the overall EA.

The communication among specialists may not always be smooth and productive. Each specialist becomes an advocate for the part of the environment that he or she is studying. This circumstance can be healthy, as it promotes debate and interaction, but it can also be negative. However, the project manager's job is to keep the debates in perspective, for example by drawing attention to indirect impacts and linkages among impacts. Advocacy positions tend to oversimplify and do not reflect the blurred and conflicting societal choices that are at the heart of project issues. Respect for technical judgements is essential: no matter how good the project manager may be, he or she cannot simultaneously be an expert in all disciplines. The dilemma is to have sufficient humanity to realise that he or she must trust the staff and have the presence of mind to challenge them. The aim must be to synthesise without diluting technical conclusions and to display and contrast objectively the difference among impacts.

Report writing

Perhaps the most important activity in the EA process is the presentation of the ES. The ES aids the decision-makers in their final decision relative to the particular project, and being a public document it will be scrutinised by interested agencies and groups. Therefore, it is critical that special care is taken in the preparation of the statement. Canter (1996) summarises from other authors five basic principles to be remembered:

◆   always have in mind the audience of the report; in the case of an ES assume that the reader is intelligent but uninformed;

◆   decide on the purpose of the report, to convey the environmental consequences of the proposed development;

◆   use simple and familiar language, since the ES requires the submission of a non-technical summary;

◆ ensure that the presentation of the report is well structured; and
◆ make the report visually attractive.

The ES should follow a logical process and be prepared in a consistent manner which can aid in communicating information to both technical and non-technical audiences.

As already noted, communication is critical in the EA process. Both informal and formal verbal presentations are typically required during the planning and execution of an EA. Such presentations are involved in scoping and public participation efforts, soliciting relevant environmental data, choosing pertinent impact prediction and assessment techniques, comparing alternatives and selecting the proposed action, and in identifying and evaluating mitigation measures. A variety of audiences may be involved, and the information presented should be tailored to specific interests and needs of such audiences. Verbal presentations must be carefully planned in terms of applying basic public–speaking principles and identifying key topics to be addressed. Examples of such topics include project need, key existing environmental resources and issues, the alternatives analysis, and key environmental benefits and costs of the proposed project. Numerous types of visual aids can be used to enhance presentation effectiveness. Finally, there is no substitute for practising presentation in order to improve the verbal communication component of EA studies.

The major written communication involves the preparation of an ES. Sound principles of technical writing should be utilised, including the development of outlines, careful documentation of data and information, the liberal use of visual display materials, and the careful review of written materials so as to ensure effective communication. Canter (1996) has developed a generic topical outline for an ES (Table 6.3).

The generic ES format given in Table 6.3 is widely employed with an element of variance to accommodate EA procedural guidelines and format modifications caused by the nature of the proposed development. A defined report structure enables the user to follow the purpose of the EA study, whereas an ill-defined structure will again lead to a breakdown in the communication process.

To further aid the communication of the study, Canter (1996) stipulates a number of useful general writing suggestions to prevent misinterpretation of the report findings and a failure in the communication process. The suggestions include: do not use clichés and catchwords; endeavour to make the ES succinct and clear, with minimal use of written texts and liberal use of visual display methods; avoid vague generalities in the ES; avoid creating a credibility gap as a result of too many technical errors and mistakes in the document; include both pro and con information with regard to a proposed action (i.e. always avoid bias); try to provide as complete a document as possible within the time frame and monetary constraints associated with a given writing effort; take care to prevent plagiarism of existing documents and avoid improper referencing; provide a document that has continuity from one section to another; avoid inconsistencies in writing and presentational style due to project team member inputs; avoid generalities in the presented information; and make sure that all presented data are included for a reason and are properly displayed and interpreted.

TABLE 6.3 Generic topical outline for an ES

I	Abstract or executive summary
II	Chapter 1: Introduction
III	Chapter 2: Delineation of need for project
IV	Chapter 3: Description of proposed project
	What is it and how will it function?
	When will it occur (timing for construction and operation)?
	Extent of effectiveness in meeting need
V	Chapter 4: Description of affected environment
	Components of baseline conditions and study area boundaries
	Interpretation of existing quality for components
VI	Chapter 5: Impacts of proposed project
	Identification of and description and/or quantification of impacts on environmental components
	Interpretation of significance of impacts
	Mitigation measures for adverse impacts
VII	Chapter 6: Evaluation of alternatives
	Description of alternatives
	Selection method and results leading to proposed action
VIII	Chapter 7: Planned environmental monitoring
	Need for monitoring
	Description of monitoring programme
	Outputs and decision points
IX	Selected references
X	Glossary of terms
XI	List of abbreviations
XII	Index
XIII	Appendices: Pertinent laws, regulations, executive orders, and policies
	Species list
	Impact calculations
	Technical description of project
	Construction specifications to mitigate negative impacts
	Description of scoping programme
	Description of public participation programme
	Environmental factors considered and deemed not relevant.

Source: Canter (1996)

When preparing an ES it is important that at least one person, preferably two, has the responsibility for reviewing the document for consistency from one chapter to another, and for consistency in terms of project information and anticipated impacts. The typical approach to the preparation of the ES will include the compilation of a draft report which is then subject to repeated internal review prior to its release to either external review or the preparation of a final draft. The presence of a technical expert in the review process will certainly enhance the production of the report and minimise the possibility of errors. A person unfamiliar with the project in the review process may additionally help to fashion the document in a way to enhance further the communication of information. Each technical aspect of the EA may require a different reviewer.

The project manager must remember that factual accuracy is indeed not a luxury. It is essential. The possibility exists that proponents, reviewers and bosses may tone down or even omit impacts from the ES. The project manager has the obligation to prevent this, even though he or she may receive pressure to do otherwise. Factual accuracy is typically threatened more subtly by sloppy communication than by outright attempts to deceive or soften impacts. It is easy to refer to impacts as severe, adverse, significant, important, negligible, moderately adverse or minimal. Without a pre-defined standard of measurement for such terms, they are meaningless and misleading. Many project managers have succumbed to the over-willingness to accommodate their client, thus compromising their professional integrity.

The application of quality assurance (QA) procedures will certainly improve the technical management and preparation of the final ES. The procedures will help focus the project team's attention on the delivery of a product that meets the requirements of the employed EA procedures and the fiscal budget of the study and satisfies the client's needs.

Financial control

One of the uncertainties relative to the planning and conduct of an EA is related to appropriate costs for the study. There are no systematically developed cost algorithms which could be used for estimation purposes. Unfortunately there is a never-ending call for the study to be done as cheaply as possible. Such an approach can prove costly in the long term if the development is called before a planning inquiry and the costs of legal fees are encountered. It is undoubtedly the preferred option to produce a quality product but the project manager is inevitably forced to focus on the realities of producing a good ES within the budget provided. It is a disappointment that this constraint is seldom taken into consideration when the ES is reviewed by a third party.

The widely held belief that the EA is an expensive part of the project management cycle of the development is false (Table 6.4). As a general rule, it is fair to say that the costs of an EA will reflect the complexity of the development, the nature and range of environmental issues to address, the logistical difficulties of conducting the studies, and the extent and quality of existing information. There are few studies which publish the costs of EA studies, but these place costs at between 0.1 per cent and 4.0 per cent of the total project investment cost. The World Bank (1991) states that costs rarely exceed 1 per cent of total capital costs. The OECD (1992) has estimated costs of between 0.1 per cent and 2.0 per cent of the total investment and a study by Neufeld (1992) (for EAs in Canada) indicated that they cost between 0.02 per cent and 3.8 per cent of total project costs.

TABLE 6.4 Cost of undertaking EAs

Project	Type of study	% EA cost to project cost
NGL plant	Safety	0.02
NGL plant	Safety	0.02
NGL plant	Safety	0.19
Oil field pipeline and terminal	EA	0.03
SNG plant	In-house EIA	0.01
Alternative road lines	EA	0.04
Alternative road lines	EA	0.02
Pump storage	EA	0.03
Water supply scheme	EA	0.03
Reservoir	EA	0.25
Gold mine	EA	0.80
Power plant	EA	0.18
Dock, jetty and oil rig service	EA (part)	0.08
Housing and recreation development	EA	0.50
Ghana Thermal	EA	0.06
Tanzania Forests	EA	0.50
Kenya Energy	EA	0.05
Guinea-Bissau	EA	0.10
Malawi Power	EA	0.08

Sources: Turnbull (1997) and World Bank (1995)

In many instances the figure is considerably less than 1 per cent. When used effectively the EA process may actually save the developer money, and therefore it is hoped that the developer will realise the advantages of the EA process and not always consider it as another cost to the project. Undoubtedly the percentage cost can vary a great deal, however, with smaller projects being on the higher side of 1 per cent. Conversely, for extremely large projects, the costs for the EA might be in the range of less than 0.1 to 0.5 per cent of the project cost. Available data to resolve this issue are limited. However, the data available are encouraging. Table 6.4 shows that the majority of EAs cost a fraction of 1 per cent of the total project cost to complete. Some case studies on the costs and benefits of conducting EA studies (for example a paper pulp mill in Indonesia and hydroelectric developments in Canada) have been described by ESSA Technologies Ltd (1994). It should be stressed that the accurate determination of the actual project costs against EA costs is difficult to gauge. However, EAs provide a means to enhance project design and improve performance (ESSA Technologies Ltd 1994) and therefore save project costs indirectly.

The costs and time required to prepare an EA will vary with the type, size and complexity of the project; the characteristics of its physical, socio-cultural and institutional settings; and the amount and quality of environmental data available

CORDAH QA — Pro-Forma Plan for Project Proposal Design Control

Date: 11/02/97

Client/job name:

Project No.

Task No.	Task description	Task fees	Non-recoverable staff costs	Travel & other expenses	Sub-contractors, Service suppliers	Equipment use and hire	Materials, Other supplies	Total
	Staff task-related costs			**'Other Costs' for each Task**				
1		£0	£0	£0	£0	£0	£0	£0
2		£0	£0	£0	£0	£0	£0	£0
3		£0	£0	£0	£0	£0	£0	£0
4		£0	£0	£0	£0	£0	£0	£0
5		£0	£0	£0	£0	£0	£0	£0
6		£0	£0	£0	£0	£0	£0	£0
7		£0	£0	£0	£0	£0	£0	£0
8		£0	£0	£0	£0	£0	£0	£0

Staff time		Project Director	Principal		Consultant	Technical assistant	Admin	Total
Recoverable		0.0	0.0		0.0	0.0	0.0	0.0
Non-		0.0	0.0		0.0	0.0	0.0	0.0
Total staff		0.0	0.0		0.0	0.0	0.0	0.0
Total non-recoverable staff costs		0.0	0.0		0.0	0.0	0.0	0.0

Project		Fees	Expenses
Proposal development		£0	£0
Project management		£0	£0
Quality assurance		£0	£0
Client admin. meetings		£0	£0
Total		£0	£0

Report production	
Reports	£0
Graphics	£0
Total	£0

Total project cost	£0
Total project price to client	£0

Accounts Department

Invoicing schedule

	weeks

1st	2nd	3rd
6th	5th	last

FIGURE 6.1   QA EA costings format for project expenditure
Source: CORDAH (1997)

(World Bank 1991 and OECD 1992). However, a review of the administration of the EA process in Canada demonstrates how the process has evolved to minimise the cost of carrying out EAs by developing a phased approach with various levels of reports, inclusion/exclusion lists, etc. which minimise the time and effort required (ESSA Technologies Ltd 1994).

One of the approaches which can be used for developing cost estimates for inclusion in a proposal for the execution of an EA is to think through, in a systematic fashion, the activities in the study execution (Canter 1991b). These activities can be further divided into cost elements including professional person–days of effort, travel costs and other related costs such as analytical costs and printing costs. Figure 6.1 shows an example of an EA costings sheet used by an environmental management consulting company (Cordah 1997).

One of the concerns relating to the fiscal control is associated with issues that might arise and cause increases in the costs of such studies. Examples of these might include:

- an extensive period of time devoted to gathering information;
- changes to project design features which may occur during the execution of the EA;
- the necessity for planning and conducting a baseline environmental monitoring programme;
- the occurrence of controversy related to the proposed project (e.g. further project meetings); and
- the identification of unique risks that might be related to project construction, commissioning, operation and/or decommissioning not previously identified.

It is the responsibility of the project manager to monitor the progress of the study against the determined budgetary constraints. Should a variation occur to the TOR then it is the manager's responsibility to identify the variance, determine additional costs and resolve financial issues prior to the commencement of additional work. Confirming variations to the original contract will avoid embarrassment at a later date when trying to resolve invoicing and payment issues with the client.

Figure 6.1 shows that fiscal control involves matching project staff and the available budget over time. Once a budget and timetable are established, the project manager must not assume that they will be followed. It is important that the project manager tracks project expenditure regularly. These procedures are not unique to EA. Projects have a nasty tendency to show low levels of expenditure at the beginning and then build to a peak. It is in the last weeks or months of a project that unanticipated expenditure seems to occur, particularly if report production is under way. Planning for contingencies and the inevitable crunch at the scheduled time for completion is the best protection (Bingham 1992). The opposite to a well-executed project is stress, under-recoveries and an irate project director.

A project manager's first EA process can be novel, exciting and stressful. It is possible to learn from it by keeping records as it proceeds or by consciously recalling the lessons it has provided. The manager should plan what he or she would have done differently. Nevertheless, a frequently overlooked aspect of EA project management is the post-mortem, or after-the-fact look with the staff to consider if manager or staff might have approached the process or its tasks differently. At that point, it is to be hoped, there is the time to prepare for the next EA. The process proceeds by building upon a foundation and subsequently modifying it as a result of having identified problems and determined ways to rectify them. It is also important to consider EAs as building blocks of data, i.e. set a goal so that data collected in one EA provide a stepping stone to the next EA. The EA process also requires that the proponent build foundations (especially important to developing countries). They may promote EA as the means to acquire environmental training, should such a project be financed by a lending or donor institution. Training of this type is likely to attract support. They may also become allied with environmental groups, with an academic or research institution, or with consulting firms. Such a partnership within a country or region and between people from developed and developing countries will be of interest to both.

CASE STUDY: PROJECT MANAGEMENT OF AN EA FOR A SEWAGE SLUDGE INCINERATOR

An EA undertaken of a proposed sewage sludge incinerator in Belfast, Northern Ireland (Aspinwall & Company 1994b) provides an example of good practice in the presentation of an ES. The study was undertaken by Aspinwall & Company in association with Ferguson & McIlveen and Binnie & Partners. The EA study took approximately 15 months to complete and involved extensive consultations and baseline studies. The ES was produced in a presentation box comprising:

- A presentation brochure – 'Your Environment – A Clearly Better Future' detailing the plans of the Department of the Environment (DoE) Northern Ireland (NI) Water Executive to implement its Sewage Sludge Disposal Strategy to achieve compliance with the EC Urban Waste Water Treatment Directive;
- The Planning Application submitted by the DoE NI Water Executive; and
- The ES, in two volumes.

The format of the ES is shown in Table 6.5. The ES was prepared in accordance with the Planning (Assessment of Environmental Effects) Regulations (Northern Ireland) 1989 to accompany a planning application for the development of a dedicated sewage sludge incinerator to be sited at the Duncrue sewage treatment works in Belfast.

Incineration of sewage sludge and screening was found to be the Best Practicable Environmental Option (BPEO) following detailed consideration of economic and environmental factors associated with potential disposal options. Figure 6.2 shows the methodology used for the assessment of sewage sludge disposal options. The

TABLE 6.5 Layout of ES for proposed sewage sludge incinerator
_____

Volume 1
Non-technical summary

Introduction
1. Background to the proposals
2. Purpose and character of the project
3. Report structure
4. Study approach
5. Consideration of alternative disposal solutions
    1. Disposal options
    2. Disposal strategies
    3. Alternative sites
    4. Northern Ireland disposal strategy
6. Legal requirements

Proposed development

1. Introduction	9. Effluent disposal
2. Incinerator capacity	10. Operation and control
3. Incinerator feedstock	11. Odour and noise
4. Disposal process	12. Site layout
5. Reception and storage	13. Appearance
6. Dewatering and drying	14. Construction
7. Combustion and heat recovery	15. Abnormal solutions
8. Flue gas treatment	16. Ash disposal

Planning policy context

1. Belfast Urban Area Plan 2001	3. A Planning Strategy for Rural Northern Ireland
2. Belfast Harbour Local Plan	4. Compliance with policy objectives

Scoping study

1. Background	3. Identification of key environmental issues
2. Consultations	

Assessment of key environmental effects

1. Introduction	6. Landscape and visual amenity
2. Air quality	7. Ecological effects
3. Water quality	8. Human effects
4. Ground conditions	9. Traffic generation
5. Noise and nuisance	10. Waste arising

Conclusions

References

TABLE 6.5 continued

Volume 2 (Appendices)

1.	Preliminary assessment of disposal options	11.	Risk assessment
2.	Environmental appraisal of disposal options	12.	Water quality data – Belfast Lough
3.	Study of alternative sites	13.	Noise impact assessment
4.	Best practicable environmental option	14.	List of businesses in the locality
5.	Incinerator feedstock	15.	Inner Belfast Lough ASSI details
6.	Ash disposal	16.	Road traffic survey data for locality
7.	Effluent disposal	17.	Climate and meteorological assessment
8.	Consultation replies	18.	Method for the determination of dioxins
9.	Air quality for the Belfast locality	19.	References
10.	Air quality impact assessment		

Note: (The format of each section was first to describe the existing baseline conditions and then detail the potential impacts associated with each environmental criterion and outline the relevant mitigation measures to minimise or prevent the impact.)

successful implementation of the proposal would ensure that the Water Executive of the DoE NI would comply with the European Community Urban Waste Water Treatment Directive (91/271/EEC) to terminate the disposal of sewage sludge at sea by the end of 1998.

The EA studied a number of potential environmental concerns which were identified and addressed as part of the ES. The conclusions of the study were that no significant environmental effects were expected to result from the operation of the development, and significant environmental benefits would result when disposal of sewage sludge at sea is stopped.

In 1995 the DoE NI Water Executive obtained planning permission for the incinerator.

Acknowledgements
The authors thank Aspinwall & Company and the DoE NI Water Executive for their kind permission in allowing them to use this case study. The comments made in the text reflect the authors' opinions only.

Questions for thought

1. What are the principal requirements of an effective project manager?
2. How would you apportion and justify resources to the component parts of an EA?
3. What criteria would you use to reconcile costs and quality in an EA?

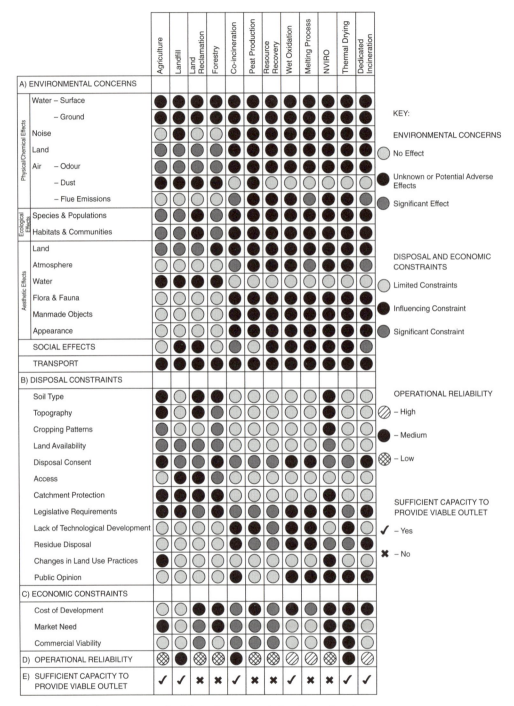

FIGURE 6.2 Assessment methodology for sewage sludge disposal options

Quality assurance
in EA

ES review and post-project analysis

◆ Introduction   128
◆ Reviewing ESs   128
◆ ES review in the UK   129
◆ International ES review procedures   134
◆ Post-project analysis: auditing and
   monitoring in EA   138
◆ Conducting a post-project analysis   139
◆ Questions for thought   146

Introduction

In EA, as in other environmental management procedures, there is a need for review mechanisms to ensure that acceptable standards of performance are being achieved and that they assist the decision-making process. Two key quality assurance components are the review of ESs and the use of monitoring and auditing studies (post-project analysis) to evaluate the approach and findings of the study, in relation to the actual effects of the development, the accuracy of predictions made and the effectiveness of mitigation measures proposed. These aspects of EA were highlighted in the international effectiveness study conducted during 1993–1996 (Sadler 1996).

Countries such as Canada and the Netherlands have established a formal ES review process to determine whether or not an ES meets its terms of reference and legal requirements. Others have adopted a less formal approach which may be inconsistently applied as well as being poorly matched by follow-up studies which seek to establish the actual effects of a project. To make a more effective contribution to sustainable development, the EA process must place greater attention on assessing the quality of the ES.

Reviewing ESs

The main purpose of the ES review is to establish the compliance of the document with the terms of reference (TOR) and legal requirements for EA. This procedural review may involve a review of drafts and the final report submitted to the decision-making authority. The actual review procedure may be referred to and be conducted by an invited panel of experts (formed of individuals with appropriate expertise related to the project) who may also make provision for public comment, as is the case in the Netherlands.

Questions relating to impartiality may arise when the authorising agency has been responsible for the EA. An independent review panel may remove any suspicion of bias in those cases where the authorising agency is an advocate for the development or holds unreasonable views against the development. The functions of the review panel may include:

◆   formulation of the terms of reference for EA;
◆   the 'scope' of the assessment, i.e. which projects should be subjected to a full or partial EA;
◆   general or specific guidelines and advice on methods of EA;

◆ ensuring that the EA has been adequately completed within the terms of reference;

◆ submitting the ES together with any separate contributions from other organisations, with recommendations to the appropriate authorising agency; and

◆ acting as a focus for the exchange of information and opinions concerning environmental affairs.

While the review process does not include measures for advising on whether or not a development should proceed, it may well help to facilitate the decision-making process. A wider scope to EA review procedure may be adopted where the quality of the information presented and the assessment procedures used are evaluated as a technical review and serve as a useful gauge of the quality of the EA process, which may go beyond an analysis of documentation produced. A technical review is an essential part of conducting a post-project analysis.

ES review in the UK

The UK system for EA does not require a formal, independent review of the ES. When an ES is submitted to the competent body for a decision it is the same body (or an independent consultant or organisation such as the Institute of Environmental Assessment) that undertakes the review. There is no formal time frame for the review and no requirement for the production of a review report. While the lack of a formal review mechanism has undoubtedly contributed to the production of some poor-quality reports (CPRE 1990, Coles and Tarling 1991), its careful use at the pre-decision stage can help in dealing with information gaps and shortcomings and set planning conditions, for example the monitoring of the project during construction/operation for a particular parameter whose impact magnitude and/or importance was not clearly established in the report.

A study of the relationship between EA and the planning system by Wood and Jones (1991) examined a sample of ESs produced within the first 18 months of the EA regulations being implemented in the UK. Local authorities were also questioned on how they dealt with ESs and any provision they made for reviewing the document. Two-thirds of the local authorities examined scrutinised the ES without the use of formal review criteria. Comments from statutory consultees in existence at the time, such as the (then) Nature Conservancy Council (NCC) and National Rivers Authority (NRA), were also utilised in addition to any in-house expertise such as from environmental health departments. Twenty per cent of authorities used consultants for the review. In one-third of cases, the local authorities found the ES to be unsatisfactory (an independent assessment of the quality by the University of Manchester found two-thirds to be unsatisfactory).

Lee (1991) suggested that 40 to 50 per cent of ESs are inadequate. In a system where the review stage is an informal process and is conducted by relatively inexperienced personnel, the quality standards required are at least inconsistent and at worst inadequate in themselves. There is, therefore, relatively little pressure on proponents of schemes to improve standards above the maximum required.

One study (Tarling 1991) compared the quality of landfill ESs for the UK with those produced under the Dutch system which incorporates a formal review by an expert body. Nearly 80 per cent of the UK ESs received were found to be unsatisfactory whereas all the Dutch ESs were considered satisfactory. While these differences in quality cannot be entirely explained by the presence or absence of a formal review procedure, it is likely to be an important contributing factor.

Environmental statement review: the Manchester review package

The Manchester review package (Lee and Colley 1990, 1992) consists of a list of criteria arranged in a hierarchical (pyramidal) structure (Figure 7.1). The review should be conducted by a team of two (or more) individuals who are sufficiently familiar with the requirements of the EA process and who ideally have technical competencies related to the particular nature of the environmental study. The team initially works independently, first by looking quickly at the whole statement to gain an overall understanding of the nature of the development, the key environmental issues, the layout of the report and the general approach adopted in the study. Each reviewer then evaluates specific aspects of the ES against the review criteria by working up through the various levels of the pyramid, starting at the base. The four review areas, each of which has categories and sub-categories for review, are:

- description of the development, the local environment, and the baseline conditions;
- identification and evaluation of key impacts;
- alternatives and mitigation of impacts; and
- communication of results.

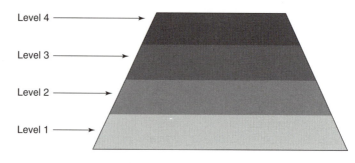

Level 4 - Overall assessment ES
Level 3 - Assessment of review areas
Level 2 - Assessment of review categories
Level 1 - Assessment of review sub-categories

FIGURE 7.1 Hierarchical structure of the Manchester ES review package
Source: Lee and Colley 1990, 1992

TABLE 7.1 ES review criteria

1. Description of the development, the local environment and the baseline conditions

1.1 Description of the development

◆   The purpose and objectives of the development should be explained.
◆   The description of the development should include the physical characteristics, scale and design as well as quantities of material needed during construction and operation.
◆   The operating experience of the operator and the process, and examples of appropriate existing plant, should be given.

1.2 Site description

◆   The area of land affected by the development should be clearly shown on a map and the different land uses of this area clearly demarcated.
◆   The affected site should be defined broadly enough to include any potential effects occurring away from the construction site (e.g. dispersal of pollutants, traffic and changes in channel capacity of water courses as a result of increased surface run-off, etc.).

1.3 Residuals

◆   The types and quantities of waste matter, energy and residual materials and the rate at which these will be produced should be estimated.
◆   The methods used to make these estimations should be clearly described, and the proposed methods of treatment for the waste and residual materials should be identified.
◆   Waste should be quantified wherever possible.

1.4 Baseline conditions

◆   A description of the environment as it is currently and as it could be expected to develop if the project were not to proceed should be given.
◆   The methods used to obtain baseline information should be clearly defined (some baseline data can be obtained from existing data sources, but some will need gathering).
◆   Baseline data should be gathered in such a way that the importance of the particular area to be affected can be placed into the context of the region or surroundings and that the effect of the proposed changes can be predicted.

2. Identification and evaluation of key impacts

2.1 Identification of impacts and method statement

◆   The methodology used to define the project specification should be clearly outlined in a method statement.
◆   The statement should include details of consultation for the preparation of the scoping report and make suitable reference to expert bodies/panels and the public consulted, guidelines, checklists, matrices, previous best practice, and examples of EAs on similar projects (whichever are appropriate).
◆   Consideration should be given to impacts which may be positive or negative, cumulative, short-term or long-term, permanent or temporary, direct or indirect.   continued . . .

TABLE 7.1 continued

♦　The logic used to identify the key impacts for investigation and for the rejection of others should be clearly explained.

♦　The impacts of the development on human beings, flora and fauna, soil, water, air, climate, landscape, material assets, cultural heritage, or their interaction, should be considered.

♦　The method statement should describe the relationships between the promoter, the planning, engineering and design teams and those responsible for the ES.

2.2 Prediction of impact magnitude

♦　The size of each impact should be determined as the predicted deviation from the baseline conditions, during the construction phase and during normal operating conditions, and in the event of an accident if the proposed development involves materials that could be harmful to the environment (including people).

♦　The information and data used to estimate the magnitude of the main impacts should be clearly described and any gaps in the required data identified.

♦　The methods used to predict impact magnitude should be described and should be appropriate to the size and importance of the projected disturbance.

♦　Estimates of impacts should be recorded in measurable quantities with ranges and/or confidence limits as appropriate.

♦　Qualitative descriptions where necessary should be as fully defined as possible (e.g. 'insignificant means not perceptible from more than 100 m distance').

2.3 Assessment of impact significance

♦　The significance of all those impacts which remain after mitigation should be assessed using the appropriate national and international quality standards where available.

♦　Where no such standards exist, the assumptions and value systems used to assess significance should be justified and the existence of opposing or contrary opinions acknowledged.

3.　Alternatives and mitigation

3.1 Alternatives

♦　Alternative sites should have been considered where these are practicable and available to be developed.

♦　The main environmental advantages and disadvantages of these should be discussed in outline, and the reasons for the final choice given.

♦　Where available, alternative processes, designs and operating conditions should have been considered at an early stage of project planning and the environmental implications of these outlined.

3.2 Mitigation

♦　All significant adverse impacts should be considered for mitigation, and specific mitigation measures put forward where practicable.

♦　Mitigation methods considered should include modification of the project, compensation and the provision of alternative facilities, and pollution control.

TABLE 7.1 continued

◆ It should be clear to what extent the mitigation methods will be effective.
◆ Where the effectiveness is uncertain or depends on assumptions about such features as operating procedures or climatic conditions, data should be introduced to justify the acceptance of these assumptions.

3.3 Commitment to mitigation

◆ Clear details of when and how the mitigation measures will be carried out should be given.
◆ When uncertainty over impact magnitude and/or effectiveness of mitigation over time exists, monitoring programmes should be proposed to enable subsequent adjustment of mitigation measures as necessary.

4. Communication of results

4.1 Presentation

◆ The report should be presented clearly with the minimal use of technical terms.
◆ An index, glossary and full references should be given and the information presented so as to be comprehensible to the non-specialist.

4.2 Balance

◆ The environmental statement should be an independent objective assessment of environmental impacts, not a best case statement for the development.
◆ Negative and positive impacts should be given equal prominence, and adverse impacts should not be disguised by euphemisms or platitudes.
◆ Predicted large negative or positive impacts should be given due emphasis.

4.3 Non-technical summary

◆ There should be a non-technical summary outlining the main conclusions and how they were reached.
◆ The summary should be comprehensive, containing at least a brief description of the project and the environment, an account of the main mitigating measures to be undertaken by the developer, and a description of any remaining or residual impacts.
◆ A brief explanation of the methods by which these data were obtained and an indication of the confidence which can be placed in them should be included.

Source: Adapted from Lee and Colley (1990)

A list of criteria, adapted from those used in the Manchester and Institute for Environmental Assessment packages, is shown in Table 7.1. The findings are recorded on a review report form, which in the case of the Manchester package, uses letters rather than numbers so as to discourage crude aggregation of the assessment criteria. Other comments, particularly examples of good or poor practice (referenced by statement section and page numbers), are also recorded. On completion of the independent analysis by the review team members, the findings should be discussed and agreement reached on the gradings to be awarded. Finally, a written report should be prepared for the requesting authority or developer. The time required to conduct the review will, of course, be dependent upon the nature and complexity of the study, the overall length of the report (which may be supported by a number of technical appendices) and the experience of the review team. A more rapid, but less comprehensive, review may be conducted by using only the higher review levels. Also, where a more focused analysis of particular themes is desired, such as how the air pollution assessment was conducted or what mechanisms were used for public participation, a selection of these criteria, together with others specially formulated, may be utilised.

International ES review procedures

ES review in Ghana

Since EA became a requirement in Ghana in 1989, a procedure has been developed which includes specific reference to the review of ESs (Environmental Protection Agency 1995). Reviews of draft ESs are conducted by a nine-member cross-sectoral technical committee, consisting of two representatives of the Environmental Protection Agency (EPA), a representative of the Ministry of Environmental Science and Technology and representatives of other governmental institutions or organisations. Additionally, relevant specialists may be co-opted on to the committee. The Ghana EIA procedure includes, as an appendix, the review criteria adopted for informal use in the UK (see Table 7.1). As part of the review process, copies of the final draft ES are made publicly available through appropriate District/Municipal/ Metropolitan Assemblies and written public comments are invited by the EPA. If there is 'a strong public concern' and the impacts are 'extensive and far reaching', the EPA is required to hold a public hearing. This is co-ordinated by a panel consisting of between three and five persons, of whom the chairman must not be resident and at least two-thirds should be resident of the area affected by the proposed development. Under these circumstances, the information received at the hearing may be used in determining the acceptability of the draft ES, after which the proponent can finalise the ES and be issued with a Provisional Environmental Permit. If the ES is not acceptable, the proponent may be required to re-submit a revised statement at a later date or to conduct further studies to modify the statement as necessary.

ES review in Canada

In contrast to the situation in the UK, an integrated technical and public review process has been developed in Canada. An expert panel is formed to undertake the review of the ES. One of the first functions of the panel is to undertake an information programme to inform the public of the nature of the review and to ensure the public is provided with adequate information to review the project. The ES is sent to any interested parties well in advance of public meetings. Written comments are also distributed prior to hearings. Advertisements are placed in the media advising of progress and procedures and the time and location of public meetings. Such meetings are held throughout the process, but in particular to comment on the ES. Any comments or issues resulting from the public meeting are taken into account by the review panel, whose members provide advice on whether the project should proceed. The comprehensive system for review used in Canada includes the provision of guidelines on conducting public participation exercises and in some cases provides for funding to enable the public to employ appropriate expertise.

Models of more formalised systems of review, based on the experience of other countries such as the Netherlands and Canada, suggest that considerable improvements in ES quality can be gained by their adoption. There are three key factors to such a system:

◆ Reviews are conducted by a specialist body. This feature of the system recognises that EA is a discipline much wider than the issues usually addressed by land-use planners. The creation of a body to deal exclusively with EA enables the members to become experts in the field and therefore set much higher standards to be attained by the proponents of the schemes. These standards need not be unrealistically high, but simply reflect national, and in some cases international, best practice standards.

◆ Reviews are undertaken by a specialist group of experts specific to each EA. This reflects the multi-disciplinary nature of EA in general and the different make-up of skills required to assess ESs for different project types and locations. Nevertheless, as in carrying out the EA, the role of the individual to manage and provide an overview of the review is crucial.

◆ The review process is separated from the decision-making process. This allows the expert body to concentrate on the adequacy of the information provided in the ES rather than be distracted by planning issues.

ES review in the Netherlands

The Dutch review system is an open process in that public views expressed on an ES (either written or voiced during a public meeting) are taken into account by an appointed independent Commission of experts, whose report on the quality and adequacy of an ES for decision making is also made public. The review is based on a series of operational criteria (Sielcken *et al.* 1996), which examine in particular the issue of alternatives. These criteria are:

◆ the quality of the description of the proposed activity (the preferred alternative) and its environmental consequences;

◆ the quality of the description of the alternative most favourable to the environment and its environmental consequences (the description of this alternative is legally required in the Netherlands);

◆ the quality of the description of other alternatives and their environmental consequences; and

◆ the quality of the comparison of alternatives (including the extent to which subjectivity has been avoided).

Three main steps are involved in this process (Scholten 1997):

1 *Step 1*: Examples of good practice and deficiencies in the ES are listed by the Commission review team taking into account the specific guidelines and general review criteria; the comments made in reviews of similar ESs; and comments made by the public and NGOs.

2 *Step 2*: Here, the operational criteria described above are used to determine if there are any crucial shortcomings, i.e. those which may influence the planning decision. These, along with the examples of good practice, are emphasised, while those deficiencies considered to be less important are either placed in appendices or left out completely so as to make the report more readable.

3 *Step 3*: This consists of the review team making recommendations to the competent authority on how and when to deal with any serious shortcomings in the ES before the proposal is approved. This may comprise a statement on any supplementary information required to make the ES adequate. Alternatively, a set of explanations and planning conditions may be provided, thereby helping to minimise any delay to decision making. Where it is considered that the shortcomings cannot easily be remedied in these ways, the review team may recommend that the deficiencies and uncertainties are monitored during project implementation and corrective measures introduced where impacts prove to be worse than predicted. Finally, the review team may judge that their experiences are sufficient for them directly to propose corrections to shortcomings in the report. This approach of providing immediate solutions has the benefit of reducing delays to the decision, but may lead some to consider that the review team is too closely allied to the developer and consultants preparing the report.

Review procedures in the Republic of South Africa

The procedure for EA in the Republic of South Africa (RSA) is known as Integrated Environmental Management (IEM) (Department of Environment Affairs 1992a). The purpose of the review component of this system is to provide an evaluation of the strengths and weaknesses of a proposal or of an assessment report submitted to the authorities (Department of Environment Affairs 1992d). On the basis of this review a decision is taken as to whether or not approval for a development proposal should be granted. The system in RSA is designed to channel development proposals down one of three routes:

1. for an 'impact assessment' when it is clear that there will be significant impacts;

2.   for an 'initial assessment' when a proposal is included in a list of activities or list of environments or where there is uncertainty as to whether the proposal may result in significant impacts; or

3.   for 'no formal assessment' where a proposal meets planning requirements and will not result in significant impacts.

All proposals following these routes have to be reviewed by the relevant decision-making authorities. There are also provisions for public and specialist reviews of reports which normally apply to the first two categories.

Authority review

The decision-making authority in receipt of a particular proposal has overall responsibility for assessing the adequacy of the document(s). The authority decides what format the review should follow (for example, whether or not public and/or specialist reviews are required), checks that it has sufficient information in order to reach its decision, and assesses the adequacy of the document using a general framework for review as guidance.

Specialist review

While the IEM procedure recognises that it is not practical to subject all reports to specialist review, this form of review may be considered for:

◆   highly technical proposals;
◆   proposals where the decision-making authority is also the proponent;
◆   proposals where the planning consultant is also the assessor; and
◆   proposals which the authority lacks the expertise to assess.

Public review

The Guidelines for Review offer considerable flexibility in how a public review should be conducted and who should be involved. For example, it could be made by the stakeholders or interested and affected parties (I & APs) themselves, their chosen representatives and/or a panel appointed by the I & APs. The format could be a public hearing or a more decentralised review by each organisation concerned in consultation with broader membership (Department of Environment Affairs 1992d). If public review is undertaken, reports and other documentation have to be made available to the I & APs.

   The review stage of the IEM system in RSA is supported by a number of Guidelines and checklists (Department of Environment Affairs 1992a, b, c, e) and is based on a framework with which to interpret the information provided by the proponent in a document and assess its adequacy. This framework is based largely on the requirements for impact assessment as described by the Department of Environment Affairs (Department of Environment Affairs 1992c).

The framework for review (Department of Environment Affairs 1992d) is constructed around a series of questions which focus on the key aspects of EA reports. In summary these are:

1. Have the principles underpinning IEM been applied?
2. Has a broad understanding of the term 'environment' been adopted in the planning and assessment?
3. Was the procedure followed in the planning and assessment process adequate for the proposal concerned?
4. Has there been sufficient consultation with interested and affected parties?
5. Is it clear where accountability for the information lies?
6. Does the initial assessment report or impact assessment report provide the necessary information in terms of the requirements of the guidelines for report?
7. Is the information in the report accurate, unbiased and credible?
8. Is adequate attention given to the reasonable alternatives identified during the scoping stage?
9. Does the report consider the possibility of cumulative impacts?
10. Are mitigating measures defined in specific and practical terms?
11. Is the information synthesised and integrated, indicating the main issues to be evaluated?
12. Are the judgements made around the issue of significance valid? Is it clear how they were made?
13. Is the information in the report communicated clearly?

Post-project analysis: auditing and monitoring in EA

EA is frequently regarded as being inadequately applied because it focuses too much on the pre-decision stage of assessment. By analysing the actual impacts of a development, the predictive ability of future EA studies can be enhanced, leading to better informed decisions and better environmental protection. This activity is described here as post-project analysis and comprises monitoring and auditing studies and the reporting of findings/recommendations.

Monitoring is used, for example, to check that any conditions imposed on the project are being enforced or to check the quality of the affected environment. Auditing is conducted after a project has become established to test the scientific accuracy of impact predictions and as a check on environmental management practices.

The term environmental auditing is used to describe a range of environmental management functions, many of which are related to the assessment of processes within existing developments and are beyond the scope of this book.

Conducting a post-project analysis

Conducting a post-project analysis may have the purpose of ensuring that terms and conditions for project approval have been implemented and that standards for environmental performance are being met. It usually includes the evaluation of baseline and post-project monitoring data and seeks to compare actual versus predicted impacts in order to assess the accuracy of predictions made in the ES and the effectiveness of management practices and procedures used. Through these activities, and by making the findings available, it is hoped that improvements can be made in project design and management and that aspects of the EA process, notably in the use of predictive techniques, recommendations and methods for impact mitigation, will be improved, making the EA process more effective (Sadler 1996). Where post-project analysis identifies unacceptable levels of environmental performance or compliance with planning consents or legislation, recommendations should be proposed to rectify matters. These would include modifications to management plans, mitigation measures, monitoring plans and so on.

Impact monitoring

The aim of impact monitoring is to determine whether or not an impact has been caused by some aspect of a development and to estimate its magnitude. The focus may be temporal (comparison of pre-occupational conditions with those during and/or after the operational period of the project), spatial (comparison of parameters at project and other 'unaffected' locations which are, as far as possible, similar in all other important respects), or both. The purpose of the monitoring programme should be decided in advance and should include selection of variables, the methodology of monitoring, statistical analysis techniques (e.g. test of significance, etc.) to be used and the numbers/distribution of samples and frequency of sampling. Without this forward planning, actually attributing an impact to the development rather than some other introduced factor or to natural variation in the environment will be difficult. The procedures for monitoring should be clearly stated and consistently followed. This is particularly important where monitoring programmes continue for several years and employ different personnel. Useful sources of information on monitoring programmes are provided in Goldsmith (1991).

Difficulties with post-project analysis studies

There are a number of factors which have made post-project analysis studies difficult to commission and conduct. The fact that post-project analysis is often excluded from the formal requirements of many EA systems in existence around the world means that developers are reluctant to commission studies, not only because of consideration of the costs involved, but also because of concerns over releasing what may be considered to be commercially sensitive information. Those studies which have been produced tend to be for research purposes, either by educational institutions or environmental agencies wishing to evaluate aspects of their EA system. Even where these institutional barriers to post-project analysis are overcome, there remain other difficulties in performing scientifically based studies. Projects are rarely implemented exactly as planned (or described in the ES), the cost of monitoring large numbers of parameters is expensive, cumulative impacts cannot be linked to individual projects, and impact predictions are often difficult to audit and interpret statistically. Culhane (1987) stated that an ideal impact prediction should be 'quantified using a technically appropriate unit of measurement' and should clearly identify 'the affected populations or resources that are measured and the time at which the effect is to occur. The ES prediction should also explicitly state the significance of the impact and be qualified by an estimate of the probability of occurrence of the impact.' The reality of a majority of ESs is that impact predictions are infrequently quantified and couched in vague language making auditing a difficult, if not impossible, task. This point is developed further in the case studies at the end of the chapter, which include a more informal approach known as 'impacts-backwards auditing' (Wilson 1995).

Before a post-project analysis can be carried out, a number of preconditions should be met. First, the development for which the ES has been written should have been operational for a sufficient time for impacts to have occurred. Second, the analysis should contain impact predictions which have been made in the ES, and third, there should be a monitoring scheme in place for a sufficient period of time for it to yield information with which to compare actual and predicted effects.

Provided that these preconditions are met, the next stage is to scope the audit, that is to determine the extent and selection of issues to be covered. This requires answers to the following questions:

- Will the audit focus on testing predictive techniques?
- Will it focus on impacts of special interest/concern, such as health impacts or water quality impacts?
- Will the audit assess mitigation measures implemented and determine their effectiveness?
- Will the audit assess monitoring procedures implemented and determine their effectiveness?
- Is there some other focus of attention?

The procedure adopted in conducting the audit should be capable of relating predicted to actual effects in a scientific manner and be capable of analysing the causes of variation between them in order to determine the reasons for success or shortfall. Again, this raises questions such as:

- Is quantitative comparison between predicted and actual impacts possible?
- What is the quality/quantity of information already existing on predicted impacts?
- Does information exist already with which to test predictions?
- What is the nature and source of data to be collected?

◆ Are methods, sampling and statistical/analytical techniques consistent with those used previously in the impact predictions?

◆ Is monitoring and assessment of accuracy feasible?

◆ Are statistical tests applicable, or is some other rating system suitable?

◆ What is the significance of the predicted impacts?

◆ What is the value of the information to be gained when compared with the costs of the audit?

Finally, in order to facilitate improvements in the EA process, a report should be prepared and made available to other EA practitioners.

CASE STUDY: AUDIT STUDY OF FOUR DEVELOPMENTS IN THE UK (CEMP 1983)

This study considered four development projects in the UK: two oil terminals (Sullom Voe and Flotta), a steelworks (Redcar) and a reservoir (Cow Green). EA reports or similar documents were analysed to identify all environmental impact predictions. In addition, information was collected on impacts which had occurred, but which were not predicted. Abstraction of impact predictions was a complex and time-consuming task. Each prediction had to be evaluated to determine those which were repetitious and arrive at a single impact prediction. In some cases it was feasible to reduce the number of impact predictions by combining those concerning a particular event into a single statement. Table 7.2 presents the total number of predictions identified for the four case studies.

TABLE 7.2 Number of impact predictions for each case study

	Sullom Voe	Flotta	Redcar	Cow Green
No. of predictions	52	459	220	60

There was a considerable difference in the number of impact predictions identified. Comparing the scope of the EA studies, it was noticed that those for Flotta and Redcar were more comprehensive than that for Sullom Voe, thereby accounting for the larger number of predictions. Additionally, the reports for both Redcar and Flotta contained techniques for predicting noise/air pollution impacts and oil spill behaviour respectively. These techniques produced a large number of individual predictions concerning levels of various air pollutants at different locations (Redcar) and the behaviour of oil spills in a variety of environmental conditions (Flotta). The number of such predictions accounted for much of the difference in numbers of predictions between Flotta/Redcar and Cow Green/Sullom Voe.

In some cases it was not possible to audit some predictions due to the form of presentation, changes in the design of the projects which might render certain predictions obsolete and irrelevant, and the fact that a number of predictions were contingent on certain assumptions concerning environmental conditions (e.g. windspeed).

The major finding was the low number of predictions audited for Flotta and Redcar. The numbers audited constitute 3.7 per cent and 9.5 per cent of the total number of predictions for Flotta and Redcar respectively. Although the proportions for Sullom Voe and Cow Green were significantly greater at 52 per cent and 48 per cent respectively, the figures show a far from satisfactory state of affairs since the Sullom Voe and Cow Green projects have the most intensive and varied monitoring programme of the four case studies. Auditing of predictions sometimes led to a situation in which it was impossible to come to a firm conclusion regarding the accuracy of predictions. This was due to the data/information only giving an indication of accuracy.

It was possible to audit only 8 per cent of all predictions. Of those audited it was possible to come to a firm conclusion on their accuracy for 82 per cent of the predictions. In the Redcar, Flotta and Cow Green cases, approximately 50 per cent of the predictions were accurate. The Sullom Voe audit showed that 66 per cent of the predictions were accurate and 34 per cent inaccurate. Although an attempt was made to determine the extent to which inaccurate predictions under- or over-estimated actual impacts, it is impossible to come to any conclusion covering the case studies. In the Flotta and Sullom Voe ESs there were insufficient inaccurate impact predictions which could be classified. The results from the Redcar and Cow Green cases showed opposite situations. In the Redcar case study it was found that five predictions over-estimated actual impacts and one under-estimated the impact. The Cow Green case study reversed this finding, however, with seven predictions under-estimating impacts and two predictions over-estimating impacts.

Very few predictive techniques were identified and it was difficult to test those utilised. It was found that audited predictions were not based on use of a technique and that when a technique was identified (Redcar) it was not possible to audit the predictions. Therefore no conclusions could be drawn on utility of specific techniques.

The main conclusion of the research, in terms of testing predictions, was that it had been very difficult to audit the impacts predicted for developments. Impact predictions were not phrased in a way which allowed auditing, and they can become obsolete very easily. In addition, existing monitoring programmes were not very useful in providing data to allow predictions to be tested in a scientifically acceptable manner. The number of impacts audited was so small that generalisations on 'best' techniques and methods were impossible.

CASE STUDY: THE GREATER MANCHESTER METROLINK SCHEME

The Metrolink is a light rapid transit (LRT) scheme which has been in operation since 1992. It was constructed to provide a north–south link across the city of Manchester, UK, and includes a 'street running' tram system with overhead power supply with a

Impact type	Scheme Feature Causing Impact											
	Direct Impacts of Scheme						Indirect Impacts of Scheme					
	Construction Stage			Operating Stage			Construction Stage			Operating Stage		
	A	B	C	A	B	C	A	B	C	A	B	C
Accidents	0	0	0	0	0	0	×/0	×/0	×/0	0/+1	0/+1	0/+1
Severance	0	0	0/−1 (−1)	0	0	−2 / 0/−1	×/0	×/0	×/0/−1	0/+1	0/+1	0/+1
Noise and vibration	0	0	−	−2,0/+1	−2,+1	0/−1	×/0	×/0	×/0	0/+1	0/+1	0
Energy use	0	0	0/−1	0	0	0	×/0	×/0	×/0	0/+1	0/+1	0/+1
Air quality	0	0	0	0	0	0	×/0	×/0	×/0	0/+1	0/+1	0/+1
Visual impact	0	0/−1	−	0/−1	−1	+2 / +1	×/0	×/0	×/0	0	0	0
Historic and archaeological features	×	×	0	×	×	×	×/0	×/0	×/0	×	×	×
Nature conservation	×	×	×	×	×	×	×/0	×/0	×/0	×	×	×
Other impacts	×	×	×	×	×	×	×/0	×/0	×/0	×	×	×

A: Altrincham – City Centre Route   B: Bury – City Centre Route   C: City Centre Route

1983 Prediction

Actual impact 1992

(−1) Key Impact

−1 : Prediction inaccurate in 1983 evaluation

Summary Notation for Completion of Assessment Matrix for the Greater Manchester Metrolink (Phase 1)

+2 A significant favourable impact, *could be* of sufficient importance by itself to be a significant influence on the acceptance of the scheme.

+1 A small favourable impact, *unlikely* to be of sufficient importance by itself to be a significant influence on the acceptance of the scheme.

0 Zero or very small favourable or unfavourable impact, *very unlikely* to be of sufficient importance to be a significant influence on the acceptance of the scheme.

−1 A small unfavourable impact, *unlikely* to be of sufficient importance by itself to be a significant influence on the acceptance of the scheme. Ameliorating measures may nevertheless be desirable provided these are cost effective.

−2 A significant unfavourable impact, which *could be* of sufficient importance by itself (if no ameliorating measures are taken) to be a significant influence on the acceptance of the scheme. Ameliorating measures are likely to be desirable.

× Impact not assessed.

FIGURE 7.2 Predicted and actual impacts associated with the construction and operation of the Greater Manchester Metrolink (Phase 1)
Source: Jones and Lee (1993)

link between two major railway stations, Piccadilly and Victoria. Extending out from the city centre, the LRT runs along an existing local rail network to Bury in the north and Altrincham in the south. Possible future extensions to this first phase of the LRT are planned.

A post-project audit of the Metrolink (Phase 1) was carried out between May and November 1992 in order to determine the positive and negative environmental impacts resulting from the construction and operation of Phase 1, and to compare the actual and predicted impacts of the scheme (Jones and Lee 1993). The EA study on which the audit is based was produced in 1983 (Lee *et al.* 1983) and was a non-statutory, strategic-level assessment covering a range of proposed rail routes including the Phase 1 Metrolink. No specific project level study was performed as legislation did not come into force until 1988 in the UK.

A matrix approach to the assessment was used. The predicted and actual impacts were analysed and any inconsistencies noted. Key environmental impacts were identified as those being scored as 1 or 2 (Figure 7.2).

While there were relatively few significant impacts, the most important ones were considered to be the direct severance impacts in the city centre during the construction stage, and the direct visual impacts in the city centre during the construction and operating stages.

The accuracy of predictions was assessed according to the differences between scores for predicted and actual impacts. When this was less than one, the prediction was considered to be 'reasonably accurate'. This was the case for most of the predictions. Existing data were generally used to assess the actual impacts using a variety of sources, and these were not always consistent with the data used in making the predictions in the 1983 study. Given also that only the initial stages of operation (i.e. until November 1992) have been examined, the audit findings should be treated as provisional.

CASE STUDY: IMPACTS-BACKWARDS AUDIT OF A SMALL SURFACE COAL MINE (WILSON 1995)

A project prepared for the US Environmental Protection Agency (EPA) in 1992/93 by Lee Wilson and Associates, Inc., included the design and testing of a practical approach to post-project analysis called 'impacts-backwards auditing' (Lee Wilson and Associates, Inc. 1992; 1993; 1995). This relies upon reports of actual, observable impacts after a project has been constructed and checks impacts arising against those predicted in the ES. The procedure was tested at mining sites, including a small strip mine in Oklahoma where the EPA had completed an EIA study two years previously. The EPA determined a 'finding of no significant impacts' (FONSI), a decision which was taken largely on the basis of mitigation measures specified by the mining company. The focus of the study was to assess how successful these mitigation measures had actually been. A nine-step procedure was developed, which is described below in the context of the case study. Lee Wilson and Associates, Inc., describe these steps more generally for application to other projects:

1. *Select project EAs to audit*
    Focus the study and limit costs by selecting a few representative projects which have operated long enough to have caused actual impacts, and for which at least some post-project information is available.

2. *Identify likely project impacts*
    Conduct literature searches and arrange meetings/discussions with stakeholders (agency personnel, local governments and local interest groups) who are familiar with the effects arising from the project or similar projects.

3. *Determine if EA may have under-predicted impacts*
    Review the EA to identify potential mis-predictions or mitigation failures. These may include impacts which differ from those described in the report, those impacts where the report noted uncertainty and those predicted using methods subsequently considered to be suspect. This step, together with step 4, is intended to reduce the scope of the audit.

4. *Determine priorities for impacts requiring further investigation*
    Use selection criteria such as the following to help decide priorities: magnitude of apparent error (especially if under-predicted); relative importance of the impact; level of public controversy and/or scientific uncertainty; the resources and effort required to conduct the study.

5. *Prepare approach for field investigation*
    Develop a detailed study plan to evaluate each impact selected in step 4.

6. *Identify actual project impacts*
    Conduct field studies and hold meetings/discussions to determine what actually happened to resources in the area of the project, including identification of cause–effect relationships possibly accounting for the impact.

7. *Comparisons with the EA report*
    Assess the EA report and determine if a mis-prediction actually occurred. In this step, the EA is presumed correct unless there is evidence to say otherwise. The final two steps apply to actual mis-predictions.

8. *Determine cause of error*
    Determine and explain why a particular prediction was not correct. Typically, this step requires determining why an impact did occur (cause–effect analysis). Some errors may be the result of poor data gathering and/or interpretation, poor methods and/or poor use of good methods. Other errors may be beyond the control of the EA team, for example when a project changed its design after the EA was completed, or the cause of the impact was not related to the project under investigation.

9. *Apply the lessons learned*
    Use the results of the audit to help inform the EA process for other similar projects.

The project selected for the study included both active operational and post-operational/reclaimed areas so that different stages in the mine development could be considered. Twenty representative impacts of small surface mines were chosen, based on the contents of the original EA study, discussions with experts and a literature review. These included three impact categories: impacts involving off-site public health and socio-economic effects of active mining operations; operational impacts on hydrology, water supply, water quality and land; and impacts related to the success of the reclamation programme such as vegetation productivity and changes to habitat type and quality (this final category was given the highest priority in the study).

A detailed study plan was devised to evaluate impacts in the priority category, which included: observing the local conditions; interviewing regulatory personnel, local officials and neighbours; analysing monitoring data, especially water quality and water-level records for the site; reviewing regulatory files and environmental data bases; and some aerial photograph interpretation. Impacts associated with both on-site (for example, equipment operation and blasting) and off-site (coal transport and storage) operations were evaluated and compared against the original ES. The findings were in agreement with the EA study in that there were no significant impacts from the coal mine. However, it was felt that there were several predictions made in the ES concerned with effects on neighbours (dust, noise, traffic and blasting) and the cumulative loss of habitat which should have been described as 'minor impacts' or 'potentially significant impact to be minimised by mitigation' rather than stated as having 'no impact'. In explaining why there were minor errors in the predictions made, four factors were proposed: (1) because the project was small, the EPA had invested relatively few resources in quality control of the EA; (2) there had been no impact scoping; (3) little consideration had been given to cumulative impacts; and (4) for many impacts, mitigation had been assumed to be more effective than it actually was.

The final, and most important, step in the impacts-backwards approach is to report on the lessons learned from the audit so that future EAs for similar projects may be improved. In this particular case, the main lessons were: (1) that the audit was an appropriate method to provide the necessary quality control for the EA process; (2) that simple scoring of impacts should be included in the EA to help identify those project-specific or site-specific conditions requiring special attention, and to identify cumulative impacts; and (3) that future EAs should be more realistic with respect to mitigation and that they should give greater consideration to plausible worst-case scenarios. As a result of the audit study, the US EPA placed a greater burden on similar coal-mining project applicants to demonstrate that impacts will be mitigated to levels which are insignificant.

Questions for thought

1. How can monitoring be used in EAs?
2. How important is post-project monitoring to EA impact prediction verification?
3. What is the difference between auditing and monitoring?

4. Obtain an ES for a proposed development project and apply an ES review technique as described in the chapter.
5. Discuss the merits of and distinguish between technical and procedural review techniques.

Strategic Environmental Assessment

◆ Introduction   150
◆ Benefits of SEA   150
◆ Assessment of cumulative effects   151
◆ Comparison of SEA and project-level EA   152
◆ Key tasks and activities   155
◆ Using SEA   157
◆ Questions for thought   169

Introduction

Since its inception in the early 1970s, EA has been largely applied to project authorisation and therefore occurred late in the planning process. Despite the utility of project-level EA, such as improvements in project design and planning, there have been deficiencies which arise from focusing only at this level of the planning process. Examples of difficulties include the assessment of indirect and cumulative impacts and a lack of detailed analysis of project alternatives because they were ruled out at an early stage of planning. Sustainable development objectives cannot be achieved through this piecemeal approach to EA and there is a growing recognition that EA could be used to greater advantage if it were utilised earlier than in the project proposal stage, that is, in the evaluation of the likely significant environmental consequences of a policy, plan and programme. This process has been variously described as strategic environmental assessment (SEA), environmental appraisal and programmatic EA.

Benefits of SEA

To date, substantial experience of using SEA has been limited to a few countries such as Australia (Sippe 1994), Canada (Le Blanc and Fisher 1994), the Netherlands (Ministry of Housing, Spatial Planning and the Environment 1989, 1992, 1994), New Zealand (Dixon 1994, Gow 1994) and the US (Webb and Sigal 1992), and aid agencies such as the World Bank (DHV Environment and Infrastructure 1994, World Bank 1993, 1995b). However, there is now an emerging interest in introducing this form of EA, for example in the evaluation of local government development plans in the UK (Department of the Environment 1991, 1993, 1994b). Elsewhere, many countries are creating new EA systems or modifying existing ones so that they can embrace the concepts of SEA, for example Hong Kong (Nair *et al.* 1994, Au 1994, Law 1994) and Namibia (Directorate of Environmental Affairs 1995).

SEA can help introduce consideration of environmental issues at an earlier stage in the planning process and thereby contribute to the formulation of environmentally sustainable policies. Other benefits include:

◆ SEA encourages consideration of alternatives which may be ruled out or ignored in project-level EA;

◆ it assists in selecting appropriate sites for projects subsequently subject to EA;

◆ potential environmental problems may be anticipated earlier, thus facilitating long-range environmental planning;

◆ the assessment of cumulative, indirect, synergistic, delayed, regional, transboundary or global impacts is more effective (see below);

◆　the time and effort required for project EA is reduced by identifying issues, initiating baseline studies and assembling data at an earlier stage (by implementing SEA, some project EAs may not be needed); and

◆　the assessment of the environmental effects of policies which may not be translated into specific projects is made possible.

Assessment of cumulative effects

One of the most significant limitations in project-level EA has been in the assessment of the effects of a number of different developments within the same geographical area or economic sector. Project-level EA systems have tended to focus on specifically defined types of project which by virtue of their size and location are considered likely to have significant environmental effects. This approach has the disadvantage that many small individual projects, which in themselves have relatively minor impacts, are not the subject of EA, even though their collective effects may be significant. The established use of EA has also tended to exclude those impacts which are distant (in time or space) from the project itself, for example the additive or synergistic effects arising from existing or other proposed developments. Additive effects generally refer to situations in which the magnitude of an impact is directly proportional to its size, whereas synergistic effects arise when the resulting environmental impact is greater than the sum of its constituent inputs. Project-level EAs have also often ignored those secondary developments which may arise as a result of an initial project, for example the proliferation of retail and housing developments which follow on from the opening of a new road. The process of cumulative effects assessment (CEA) seeks to address these deficiencies and, with the exception of the analysis of the cumulative impacts of a single project, is more appropriately performed within the scope of an SEA.

Cumulative effects take a variety of forms: frequent and repetitive or high density impacts on a single environmental medium; synergistic effects from multiple sources on a single environmental medium; impacts resulting some distance from the source; and secondary impacts resulting from a primary activity. CEA methods attempt to analyse and predict the potential for a range of effects, accumulating from actions over space and time, using techniques such as matrices, networks, causal analysis, and systems modelling. The identification of appropriate spatial and temporal scales and the identification of suitable thresholds/ecosystem carrying capacity is central to assessing cumulative impacts. Ecosystems rather than administrative boundaries are the most appropriate spatial scale. Although the effect of an individual project is negligible, or can be made negligible through mitigation, it is when the combined effects of several projects exceed a threshold that significant environmental damage may become evident. Either an incremental approach or a comprehensive approach to the development may be used to prevent thresholds being exceeded.

Cada and Hunsaker (1990) have described a cumulative impact assessment for a hydropower development in Canada. Figure 8.1 shows four different pathways which lead to cumulative environmental effects. Pathways 1 and 2 result from the effects of a

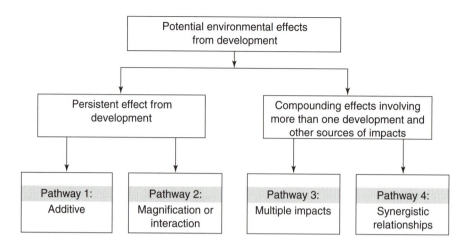

FIGURE 8.1   Potential pathways leading to cumulative environmental impacts
Source: Cada and Hunsaker 1990. Reprinted by permission of Blackwell Science, Inc.

single project on a resource or ecological receptor. Pathway 1 represents simple additive effects such as changes in water temperature, rates of erosion or habitat loss, i.e. many of the direct effects predicted by project-level EA. In many cases such impacts are not cumulative unless ecosystem capacity is exceeded. Pathway 2 represents interactions which can occur between impacts arising from a single development. For example, compared with natural water bodies, deep storage reservoirs often have increased temperature regimes, lower dissolved oxygen and contain trapped contaminants, such as heavy metals, adsorbed onto sediments. Each of these factors can affect aquatic biota.

Figure 8.1 also shows the cumulative effects of more than one development or other source of stress. Pathway 3 occurs when impacts arising from multiple developments give rise to multiple environmental effects which are additive by nature. This is similar to Pathway 1, except that impacts are identified from multiple rather than single developments. Pathway 4 indicates that impacts arising from multiple developments give rise to synergistic/interactive effects, i.e. the combined effects are greater than the sum of individual effects. This is similar to Pathway 2, except that impacts are identified from multiple rather than single developments.

Comparison of SEA and project-level EA

Project-level EA and SEA have a common set of objectives and should relate closely to each other within the same policy and planning process (Lee and Walsh 1992). SEA is applied at the earlier stages or tiers of planning action (policies, plans and programmes) in sectors such as transport, energy, tourism and land use and regional development plans. This is in advance of (or possibly instead of) EA studies applied to

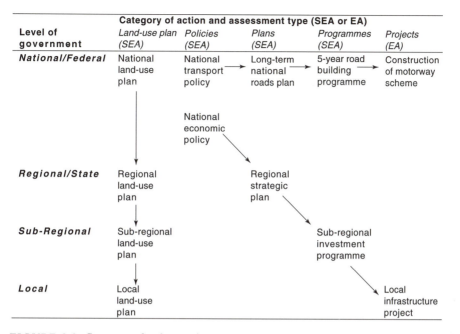

Level of government	Category of action and assessment type (SEA or EA)				
	Land-use plan (SEA)	Policies (SEA)	Plans (SEA)	Programmes (SEA)	Projects (EA)
National/Federal	National land-use plan	National transport policy	Long-term national roads plan	5-year road building programme	Construction of motorway scheme
		National economic policy			
Regional/State	Regional land-use plan		Regional strategic plan		
Sub-Regional	Sub-regional land-use plan			Sub-regional investment programme	
Local	Local land-use plan				Local infrastructure project

FIGURE 8.2 Sequence of actions and assessments (SEA and EA) within a comprehensive planning and assessment system
Source: Based on Commission of the European Communities (1990)

projects. This relationship is illustrated in Figure 8.2, which shows the example of transport planning. National transport policy is established at the highest tier and is likely to require the broadest, least detailed form of SEA. This policy then provides the framework within which a national roads plan is developed which in turn leads to a five-year road-building programme and finally to the approval and construction of specific motorway schemes.

Lee and Walsh (1992) have described the procedural and methodological similarities and differences between SEA and project-level EA. Assessments are conducted in much the same way for SEA as for EA, with respect to screening (guidelines such as positive/negative screening lists can be used for both processes) and scoping (in both cases there is a need for the development action proponent to decide on the range of issues and level of detail required, in consultation with others). Similarly, a strategic environmental assessment report (SEAR) should include the kind of information to be expected in a project-level ES.

The SEAR or ES is made publicly available for information and comments. The products of these consultations, which may be more extensive due to the more strategic nature of information and issues dealt with in a SEAR, are made available to the competent authority and used in its decision making. Evidence from the SEAR/ES and consultation documents is then used by the competent authority in making its decision on the action.

These differences between EA and SEA have been described by Lee and Walsh (1992) and are summarised below:

- Differences in scale: The scale of an SEA tends to be much greater than that of a project-level EA, first because the proposed development action is likely to contain a number of different activities compared with a single project. Second, the range of alternatives that may be considered is likely to be greater (for instance, including alternative locations, technologies, and land-use patterns). Third, the range of environmental impacts to be assessed and the area to cover may be greater.

- Differences in timing: The time interval between planning and approving an action and the implementation of the specific activities which give rise to environmental impacts is greater in SEA than in project-level EA. This can result in more uncertainty in impact predictions for SEA since less may be known about the eventual action which may also change as it goes through the planning process.

- Differences in available time: The time available for gathering and analysing information for an SEA is, with the important exception of some policy decisions, greater than for EA.

- Differences in degree of detail and level of accuracy in information needed: The degree of detail and the level of accuracy of information needed for policies, plans and programmes decision making is generally less than that needed for project evaluation and decision making. It will often be much less at the highest stages in the planning process.

- The procedural differences between SEA and project-level EA arise from the timing at which SEA and EA are triggered in the planning process. Lee and Walsh (1992) have documented five issues for consideration:

  - Confidentiality: The draft contents of certain policies (such as details of central government budget proposals), plans and programmes, may be considered too sensitive to release for public consultation prior to their approval. As in the case of EA, this may be handled by exemptions from certain consultation arrangements in those cases where confidentiality justifies this.

  - Constitutional issue: Certain actions (such as high-level policy decisions) are approved by national cabinets acting under conditions of collective ministerial responsibility. If these were subject to SEA law, the cabinet decisions relating to them may be subject to legal challenge in the courts. In Canada, this has been addressed by incorporating an EA procedure within federal cabinet decision-making procedures.

  - Procedural deficiencies: In order to be fully effective, SEA should be integrated into existing procedures at key decision-making points for policies, plans and programmes. These procedures should have the potential to meet SEA requirements relating to the provision of documentation by the proponent, for consultations based on this and the use of this combined information in decision making by the competent authority. The extent of existing provisions of this kind, within the earlier phases of planning processes in many countries, is variable. However, within a tiered system of EA, there is considerable flexibility in selecting the stages in the planning process at which to carry out assessments. The existence of suitable planning procedures into which these may be integrated is one of the factors in making that selection. Nevertheless, in certain cases, some institutional and procedural strengthening may still be desirable.

◆ Proponent–competent authority relationship: In certain cases, the proponent belongs to the same organisation as the competent authority. In the case of policies, plans and programmes, this is likely to occur frequently. One means of safeguarding the objectivity and quality of the EA process, in this type of situation, at the project level is to submit the ES to review by an independent environmental authority or commission. A similar kind of solution may be needed to safeguard the SEA process.

◆ Curtailment of competencies: SEA may be resisted by some government departments as an intrusion into their area of competence. In fact, SEA (like EA) is not intended to change the decision-making responsibilities of competent authorities. However, there is little doubt that the introduction of SEA, particularly at the national policy-making level, is a sensitive issue. It provides a real challenge to governments and, more particularly, departments with developmental responsibilities, to give greater meaning and credibility to their role in promoting sustainable development.

Key tasks and activities

While assessment methods for SEA and EA are broadly similar, there are differences in the scale and timing of the study as well as the degree of detail required. EA methods for impact scoping, identification, and prediction can be adapted for use in SEA alongside methods used in policy analysis and plan evaluation such as scenarios, planning balance sheets, cost benefit analysis, multi-criteria analysis and life cycle analysis (DHV Environment and Infrastructure 1994, Federal Environmental Assessment Review Office 1992, Sadler and Verheem 1996, Verheem 1994).

As a tool for evaluating sustainable development, SEA works well when a series of development scenarios are appraised in terms of their social and economic as well as environmental consequences within defined time and geographical limits (see case studies). Crucial to the success of such a study is a thorough examination of stakeholder interests, implemented by means of a public consultation and participation process. The basic steps for conducting an SEA are shown in Table 8.1.

TABLE 8.1 Some basic steps in conducting an SEA

Task	*Activities*
1. List the objectives of the policy, plan or programme, including the formal decisions that need to be taken, and identify the constraints.	1.1 State objectives and priorities. 1.2 Identify any conflicts and trade-offs between them. 1.3 Indicate how binding the constraints are and whether they will change over time or be negotiable.
2. Scope and analyse existing environmental issues, problems and protection objectives.	2.1 Focus on the main issues and problems that could be affected by the plan, policy or programme (consider both negative and positive issues). 2.2 Use relevant environmental policies to list relevant environmental protection objectives for these issues/problems.
3. Specify reasonable options for planning decisions and identify their environmental consequences.	3.1 Identify and evaluate environmental issues and impacts, including cumulative impacts and sustainability issues (do not disregard likely effects simply because they are not easily quantified).
4. Carry out a programme of public consultation and participation.	4.1 Identify key stakeholders and seek their views and concerns over the proposal and on the results of the SEA prior to any decision on implementation.
5. Identify opportunities for mitigating or compensating impacts considered by stakeholders and assessors to be significant and suggest a preferred option.	5.1 Focus on those impacts which are material to the decision. 5.2 Compare these impacts with relevant environmental protection objectives. 5.3 Make a comparison of alternative options, including the 'without proposal' alternative. 5.4 Test the sensitivity of the outcomes of the SEA to possible changes in conditions or to the use of different assumptions or development scenarios.
6. Establish a monitoring programme where necessary and decide when to evaluate the implemented policy, plan or programme.	6.1 Identify further requirements for assessment where possible. 6.2 Specify any projects or other activities that may require project-level EIA. 6.3 Indicate how monitoring results of projects will be collected and used to evaluate the implementation of the policy, plan or programme.

Source: Department of the Environment UK (1991)

Using SEA

European Union initiatives

Various EU treaties and programmes have stressed the importance of integrating environmental concerns into EC policy-making activities, thereby making a commitment to a system for policy appraisal. The EC Fifth Environmental Action Programme, 'Towards Sustainability', published in 1992, investigated European issues and established the strategy for examining them together with Community policies and their role in addressing wider, international concerns. Five key economic sectors were identified for particular attention over the next decade (industry, energy, transport, agriculture and tourism) and priorities for action were drawn up as follows:

◆ sustainable management of natural resources: soil, water, natural areas and coastal zones;
◆ integrated pollution control and prevention of waste;
◆ reduction in the consumption of non-renewable energy;
◆ improved mobility management including more efficient and environmentally rational location decisions and transport modes;
◆ coherent packages of measures to achieve improvements in environmental quality in urban areas;
◆ improvement in public health and safety, with special emphasis on industrial risk assessment and management, nuclear safety and radiation protection.

The Fifth Environmental Action Programme states that 'Given the role of achieving sustainable development it seems only logical, if not essential, to apply an assessment of the environmental implications of all relevant policies, plans and programmes.' (Commission of the European Communities 1992).

Article 130(r) of the Treaty of the European Union states that 'Environmental protection requirements must be integrated into the definition and implementation of other Community policies' (Kramer 1990). In order to achieve such integration, the European Commission adopted in June 1993 an internal communication which includes the provisions that 'all future Commission actions must be screened and environmentally assessed if they are likely to have a significant effect on the environment', and 'new legislative proposals which are likely to have a significant environmental impact must be accompanied by an environmental statement' (Norris 1996). Within the Commission, the sectoral DG responsible for putting forward a policy is also responsible for undertaking the SEA of that policy, while a special unit in DGXI has a screening role, provides technical assistance and monitors progress (Walsh 1996).

In addition to initiating an SEA system for its own policies, the European Commission drafted a proposal for an EU-wide directive on the environmental assessment of policies, plans and programmes in 1990. This proposed that policies, plans or programmes would be subject to EA where they are liable to cause significant effects on the environment before they received consent to the extent that the environmental effects concerned are likely to be inadequately assessed at other stages of the planning process (Commission of the European Communities 1990). This draft proposal has subsequently been modified in a draft directive whose major change is the omission of policies from the assessment procedure (Commission of the European Communities 1997). The justification for this, given in the explanatory memorandum, is that 'general policy decisions develop in a very flexible way and a different approach may be required to integrate environmental considerations in this process'. The new draft directive applies to plans and programmes 'which are adopted as part of the land use decision making process for the purpose of setting the framework for subsequent development consent decisions which will allow developers to proceed with projects' (Explanatory Memorandum accompanying the draft Directive). This means that plans and programmes which are not formally adopted will not be subject to the requirements of the directive. Sectors for which plans would require an SEA include: transport (including transport corridors, port facilities and airports); waste management; industry (including extraction of mineral resources); tele-communications; tourism; and energy. A procedural framework for conducting an SEA for a plan or programme is provided in the new draft directive. This would require the SEA to be carried out prior to the plan or programme being adopted, and the competent body has to consult with relevant environmental authorities and/or bodies over the scope of the assessment. Guidance is provided on what to include in the ES as follows:

◆   contents of the plan/programme and its main objectives;
◆   environmental characteristics of any area to be significantly affected by the plan/programme;
◆   any existing environmental problems relevant to the plan/programme;
◆   likely significant direct and indirect environmental effects of implementing the plan/programme on human beings, fauna, flora, soil, water, air, climate, landscape, material assets and cultural heritage;
◆   alternative ways which have been considered for achieving the objectives, and the reasons for not adopting the alternatives;
◆   measures to prevent, reduce and where possible offset significant environmental effects;
◆   difficulties encountered in carrying out the SEA; and
◆   non-technical summary of the ES.

While there is provision for the public to express an opinion on a draft plan/programme and its ES before the plan/programme can be adopted, this is not at the scoping stage of the assessment

SEA in the Netherlands

A two-tier approach to SEA has been developed in the Netherlands. An SEA process for certain sectoral policies, national and regional plans and programmes was introduced as part of the 1987 EIA Act (Verheem 1992) and a process for the assessment of other policies is being implemented (de Vries 1996). The Dutch National Environmental Policy Plan, whose ultimate aim is the attainment of sustainable development, recommends that environmental screening should be expanded to include all policies and plans not currently subject to mandatory SEA (Ministry of Housing, Spatial Planning and Environment 1989, 1992). Other recent developments, which make the Dutch SEA system one of the most comprehensive in coverage of levels of decision making and policy sectors, include SEAs at the national level for waste management plans (Verheem 1994), electricity production (DHV Environment and Infrastructure 1994, Verheem 1992), water supply (Verheem 1996), political programmes for major political parties (DHV Environment and Infrastructure 1994) and the application of SEA to selected overseas aid programmes by the Dutch Ministry of Foreign Affairs (Sadler and Verheem 1996).

SEA in the UK

The environmental appraisal of policies and plans represents the British equivalent of SEA. While no formal framework currently exists, guidelines for good practice have been produced (Bedfordshire County Council/RSPB 1996, Department of the Environment 1991, 1993) and a number of studies published (see for example the Revised Lancashire Structure Plan (Pinfield 1992); Bedfordshire County Council (Bedfordshire County Council Planning Department 1995)). Since 1990 the UK government has carried out a comprehensive review of planning guidance and has issued a series of new national, regional and minerals planning notes. These give guidance on how the planning system can work towards achieving the objectives of sustainable development. Planning Policy Guidance Note 12 (PPG 12), issued by the Department of the Environment (Department of the Environment UK 1992b), is concerned with development plans and regional planning guidance and requires that environmental appraisal should apply to all types of plans, policies and proposals. The appraisal process is seen as an integral part of the plan-making and review process, which allows for the evaluation of alternatives and is based on a quantifiable baseline of environmental quality (Department of the Environment UK 1993). Its purpose is:

- ◆ to clarify the environmental objectives for the plan;
- ◆ to understand the implications for the environment of any policy option, or interacting group of policy options;
- ◆ to enable the implications for different, wide ranging, and potentially conflicting aspects of the environment to be taken into account;
- ◆ to allow environmental matters to be considered along with economic and social factors, and so to assist in making a choice between alternative policies and proposals in a way which will secure the best outcome overall; and
- ◆ to demonstrate to users of the plan how the policies have regard to environmental matters.

Examples of the use of SEA in the UK include the Firth of Forth Transport System (Raymond 1994).

One sectoral example of where SEA is likely to be used more commonly is in transport planning. EA procedures for new major road projects in the UK were published as a manual in 1983 (Department of Transport UK 1983) but these were criticised for their lack of strategic-level EA by the government transport advisory committee (Department of Transport 1992). More recently, there has been an official acceptance of the potential use of a more strategic assessment of road developments following the publication of the Design Manual for Roads and Bridges (Volume II Environmental Assessment), which states that 'In some cases, assessment needs to cover the combined and cumulative impacts of several schemes. Consideration of longer routes or a number of related schemes together may allow a better choice of alignment and design in both environmental and traffic terms' (Department of Transport UK 1994).

Sheate (1992) described a staged approach for fitting SEA into a process of formulating an environmentally sustainable transport policy. The first stage involves the setting of objectives and targets, such as reductions in accidents, congestion, CO_2 and other emissions, fuel consumption, damage to wildlife (all or some of which may be quantified). The second stage is to identify the options for achieving these objectives, for example investment in public transport; technological innovations to promote fuel conservation and emissions control (lean burn engines, diesel/electric cars, catalytic converters); introduction of fiscal measures (carbon tax, unleaded fuel, engine capacity taxation); land-use planning (examples include encouraging integrated transport schemes, facilities for pedestrians and cyclists and the discouraging of out of town shopping centres). The next stage is the building of scenarios/models in which different combinations of options and different levels of investment are considered, and it is here that SEA should be incorporated into the policy formulation process. The SEA is conducted so that the relative environmental impacts of each scenario can be identified, assessed and evaluated. Finally, an assessment of alternatives is conducted, in which either an originally chosen set of options is selected or a new mix of land-use planning, fiscal control and technological developments deemed to have the least environmental impact is recommended.

SEA in the USA

The US National Environmental Policy Act (1970) requires that all 'major Federal actions' are subject to an EA and the production of an ES. This was subsequently interpreted by the Council on Environmental Quality (CEQ), established by NEPA, to include policies, programmes and similar actions in addition to projects. Where EA is considered necessary for such an action (other than a project), a Programmatic Environmental Impact Statement (PEIS) has to be produced, although the decision to proceed is left to the discretion of the government agency (Webb and Sigal 1992). PEISs tend to be produced for groups of federal actions that are related to one another by virtue of their location, nature of activities or stage of technological development.

Examples of studies produced have included PEISs for Forest Land and Resource Management (for Sierra National Forest), a series of hydroelectric projects in the Owens River Basin, California (where cumulative impacts were considered to be important), and the Department of Energy Environmental Restoration and Waste Management Program (ERWM). These SEAs are applied at the agency plan or programme level. SEA has not been used in the US in the development of broad government policies and it should be noted that EA is still most often applied at the project level.

The majority of PEISs in the US are prepared by the US Army Corps of Engineers (which deals mainly with flood control programmes) and the US Department of Agriculture (USDA) Forestry Service (which deals with resource management such as National Forest, wilderness, pest control and timber leasing). Case studies illustrating how SEA has been used in the US are given in Canter (1989) and Sigal and Webb (1994).

CASE STUDY: ENVIRONMENTAL APPRAISAL OF THE KENT STRUCTURE PLAN

As part of the third review of the Kent Structure Plan, an environmental appraisal of all policies and their likely impacts was performed (Kent County Council 1993). This appraisal, based on sustainable development criteria, indicates the level of sustainability of the proposed structure plan policies. In carrying out this appraisal it is recognised by the authority that it cannot ensure that all policies are sustainable in all respects. This can be attributed to the fact that the current structure plan in force and existing planning permissions will largely determine the pattern of development over the next five to ten years. Furthermore, land for new development will continue to be needed in the medium and long term outside urban areas because of increasing numbers of households, a continuing need for new job–creating development, leisure facilities and other essential needs.

However, the view of Kent County Council is that 'by adopting a policy approach now which better integrates environmental concerns with land use and transport policies, the foundations can be laid for a pattern of development, urban form, and transport which embodies the principles of environmental protection and enhancement, and energy efficiency. Indeed, it is the objective of this Third Review Structure Plan that the overall effect of the sum total of its policies reflects the concept of sustainable development' (Kent County Council 1993).

The approach adopted by Kent in its environmental appraisal was to develop a matrix which evaluated policies against their likely impact on environmental receptors (geology, soils, climate, human beings, air, water, energy, land, wildlife, landscape, townscape, and open space). In order to evaluate the Structure Plan policies in terms of their impact on the environment and their contribution to the development of the concept of sustainability, these elements were grouped into three.

- local criteria: ensuring that development is carried out in the right place and in the right manner (human beings, townscape, cultural heritage, noise, open space/access to the countryside).
- county-wide criteria: maintaining the environmental resources of Kent which includes consideration of Kent's carrying capacity/critical loads (e.g. ecology, air quality, water, land/ground pollution, landscape).
- global sustainable development: ensuring that local and countywide changes have minimum detrimental and maximum beneficial effect on the global environment (renewable natural resources, non-renewable natural resources, non-renewable energy resources, energy conservation, atmospheric global change).

CASE STUDY: ENVIRONMENTAL APPRAISAL OF THE GRAMPIAN STRUCTURE PLAN

The Grampian Structure Plan was produced as a successor to the Aberdeen Area Structure Plans and, after a process of public consultation and revision of a draft statement, was submitted to the Secretary of State for Scotland in February 1995 for approval. While modifications were awaited, a non-mandatory environmental appraisal was conducted (Grampian Regional Council 1995) taking into account government guidance (Department of the Environment UK 1993) and the experiences of a number of local government authorities in the UK. The appraisal was designed to minimise subjectivity and allow for a detailed analysis of a wide range of issues. The approach adopted was to use two multidisciplinary groups including planners, environmentalists and ecologists, with representation from Scottish Natural Heritage (SNH). One group carried out a detailed appraisal during group sessions and its initial findings were then critically assessed by a second group, whose members indicated areas requiring amendment or clarification. Finally, a joint meeting of the two groups took place to discuss the appraisal scoring and to come to an agreement over the findings.

The appraisal employed a set of fourteen defined appraisal criteria which were further sub-divided for consideration in order to provide rigour to the assessment. The criteria were:

◆ Liveability	◆ Air quality
◆ Built environment	◆ Water quality
◆ Cultural heritage	◆ Land and soil quality
◆ Access to open space	◆ Landscape quality
◆ Wildlife habitats	◆ Energy efficiency
◆ Biodiversity	◆ Renewable energy
◆ Renewable natural resources	◆ Non-renewable natural resources

The appraisal criteria were used to assess each of the main policy elements in the structure plan. Strategies, policies, proposals and recommendations were appraised either individually or, where appropriate, in groups of linked elements, wherever they were adjudged to have a direct impact on the environment using a simple six-point scoring system.

A total of 107 Grampian Structure Plan elements or groups of elements were appraised and the results presented in matrices showing the score allocated together with explanatory comments. Of these, sixty-three (58.9 per cent) had overall appraisals which were scored as positive in environmental terms and four (3.7 per cent) were marked as very positive. Twenty (18.6 per cent) of the elements or groups of elements were scored as having both positive and negative features. Nine (8.4 per cent) were scored as negative in environmental terms while only one (0.9 per cent) was scored as very negative. Those elements which scored negatively or had positive and negative impacts were scheduled to be reconsidered as part of the monitoring and review process. The environmental appraisal process was considered to be a useful exercise, particularly in the way it provided a robust approach towards scoping, the structures and wording of policies (particularly in terms of their implications for the environment) and the way it facilitated debate over policy elements. Finally, the view was that the process would, in time, strengthen the Plan by broadening the under-standing of sustainability issues and it was recommended for use by the successor authorities (Aberdeen, Aberdeenshire and Moray Councils) in the monitoring and review of the Structure Plan and in the assessment of any proposed modifications made to it by the Secretary of State for Scotland.

CASE STUDY: SEA FOR THE LAKE MYVATN AREA IN SKUTUSTADAHREPPUR, ICELAND

Lake Myvatn has an international reputation for its scenic beauty and wildlife. The lake is designated as a Ramsar Site, and attracts large numbers of tourists. In 1990 the population of the area was 515 inhabitants located in Reykjahlid, Skutustadir and other areas. Traditionally, the area is based on agriculture, with hay being the main crop. Industrial development took place between 1965 and 1977 with the construction of a diatomite plant (Kisilidjan) and a geothermal power plant (Krafla). Agriculture in the area has been in decline in the last thirty years, and is now less important than those industries in terms of employment, although there is some uncertainty about the future of the diatomite plant. Tourism has also grown in the last thirty years.

An SEA was conducted in 1993 by the National Physical Planning Agency (NPPA) in consultation with the Skutustadahreppur County Council, Nature Conservation Council and the Ministry of the Environment, as part of the devel-opment of a General Plan for the County (a land-use plan for the next twenty years) (National Physical Planning Agency 1993). Four planning alternatives were evaluated for their economic and environmental impacts (Figure 8.3). The alternatives were:

1. continuation of the present situation in which population of the area (515 in 1990) and utilisation of resources remain more or less unchanged,

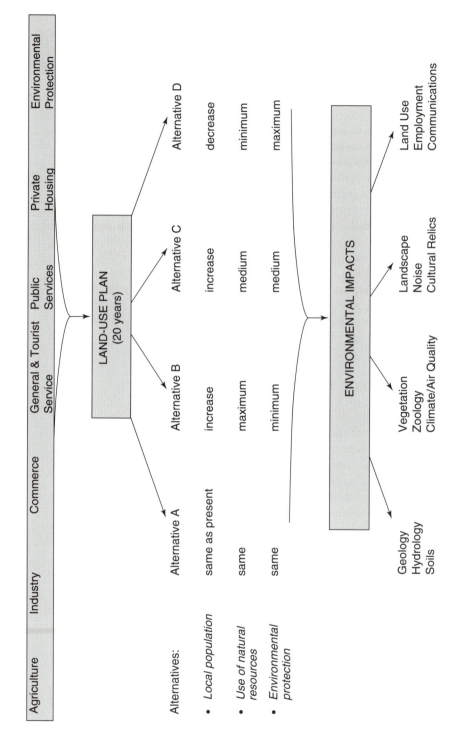

FIGURE 8.3 SEA for the Lake Myvatn area, Iceland

2. maximum utilisation of natural resources and minimum environmental protection with local population rising to 1,000 inhabitants by the year 2012;
3. medium utilisation of natural resources and medium protection, with the population reaching 700 by the year 2012;
4. minimum utilisation of natural resources and maximum protection, population reducing to 400 by the year 2012.

The environmental impact of each alternative was described in relation to: ecology, hydrology, soils, vegetation, zoology, climate and air quality, landscape and panorama, noise pollution, cultural relics, land use, employment and communications. Impacts were classified as being:

1. significant, irreversible (defined as impacts which cannot be restored by nature itself or by human efforts within twenty years of a project);
2. considerable but reversible;
3. insignificant.

Cumulative, short- and long-term impacts were described and mitigation measures for impacts proposed. For each alternative, specific details in relation to agriculture, industry, commerce, general and tourist services, public services, private housing and environmental protection were described. However, no recommendations on preferred alternatives were made. In summary, the main impacts of the four alternatives were:

Alternative A

This assumed that the development of the area continue as at present. The area could cease to be a popular tourist attraction due to market saturation and damage linked to a lack of tourist services. The loss of tourism-related revenue would lead to a loss of employment. It was recommended that a detailed plan should be established for employment and construction in order to ensure the long-term economic prosperity of the community.

Alternative B

This assumed that diatom mining from Lake Myvatn would continue, thus maintaining a source of employment and income for the local population. Such operations could, in the long term, result in a decline in the lake's biological condition, which in turn would decrease tourism, and therefore employment. As these extracted resources became depleted, a loss of employment opportunities and a reduction in local population could result. A major increase in the number of summerhouses in the area was assumed as well as a major population increase. A growth in industrial and agricultural development would create greater employment opportunities, resulting in an even greater population increase, increased threat of accidents and environmental pollution, and a temporary lack of public services, given that their organisation requires more time. The local schools would be over-subscribed and there would be an initial shortage of kindergartens and day-care centres.

Alternative C

This assumed that there would be economic growth in the area, creating employment opportunities, but due to the heavy reliance on diatom mining, the number of jobs would decline as this operation closed. Protection and soil conservation measures would require capital and manpower, giving an economic return later in the form of improved land having a greater carrying capacity for livestock and a larger population.

Alternative D

This assumed that increased land protection would decrease options on other land uses, hence job opportunities would be fewer. These short-term effects would result in a reduced population, less traffic and fewer accidents, and eventually in a diminished provision of public services. Protection measures would, however, in the long term result in increased vegetation and wildlife. The threat of cultural relics being damaged would decrease through registration and increased protection. Such measures could make the area even more attractive in the future for tourism.

CASE STUDY: SEA FOR THE VICTORIA FALLS AREA, ZAMBIA/ZIMBABWE

Victoria Falls was declared a World Heritage Site by UNESCO in 1989 at the request of the governments of Zambia and Zimbabwe. The Falls are one of the world's most scenic natural features, extending 1708 m wide and 103 m deep at their highest point. Following the rainy season, the waters of the Zambezi flow over the Falls at a rate of 500 million litres per minute. The spray cloud which is formed has led to the creation of a magnificent area of rainforest (Phillipson 1990) and gives the Falls their African name of Mosi-oa-Tunya ('the smoke that thunders'). Within the boundaries of the World Heritage Site are three National Parks (Mosi-oa-Tunya on the Zambian side of the Falls, and Zambezi and Victoria Falls on the Zimbabwean side) and the two towns of Livingstone, Zambia (population 95,000) and Victoria Falls, Zimbabwe (population 25,000). A number of rural communities exist on both sides of the border.

The tourism value of the Falls is significant to the economies of both countries, since a majority of visitors to the countries spend some of their time in the area, and this interest is growing. In Zimbabwe, there has been a fourfold increase in visitors to the area in the last ten years and an associated growth in hotels and lodges, tourism-related facilities (such as white water rafting, helicopter and light aircraft flights over the Falls, cruises on the River Zambezi) and other associated infrastructure. Tourism-related growth on the Zambian side has not been as significant, although there are some tourist hotels and lodges and other facilities around Livingstone. While there is an undoubted desire to increase the economic benefits of an expanded tourism industry in the Falls, there is also an acceptance that tourism developments should be reconciled with protecting the quality of the environment.

Following a report into tourism and environment issues and recommendations for environmental management on the Zimbabwe side of the Falls (Department of Natural Resources 1994), the governments of Zambia and Zimbabwe collaborated on the preparation of a tourism development master plan to seek sustainable development in the Victoria Falls area. An SEA was conducted to establish the likely significant, cumulative impacts of current and expected developments affecting the area and to provide information and recommendations for use in this master plan. The SEA began in November 1994 with a scoping workshop (IUCN Regional Office for Southern Africa 1994) held in Livingstone and attended by representatives of all main stakeholders in the area. This meeting established that the study would be confined to an area of 30 km radius around the Falls and would assess impacts, including cumulative impacts for a series of development scenarios over a period of ten years (1995–2005). A joint Zambia–Zimbabwe team was selected, including: sociologists; urban and natural resource planners; hydrologists; ecologists; environmental economists; and specialists in archaeology/cultural resources, aircraft noise and tourism. A steering committee consisting of representatives from the agencies on both sides of the Falls (the Zambian National Heritage Conservation Commission and Zimbabwean Department of Natural Resources) was formed and co-ordination was provided by the World Conservation Union, Regional Office for Southern Africa (IUCN-ROSA) (Meynell 1997). In summary, the main findings of the scoping workshop (IUCN Regional Office for Southern Africa 1994) were as follows:

Main adverse environmental trends:

◆ increasing noise levels;
◆ reduction of wildlife corridors;
◆ harassment of tourists by hawkers;
◆ deforestation and other removal of vegetation;
◆ encroachment of development into protected areas (especially in Zambia);
◆ anti-social behaviour;
◆ cultural 'erosion';
◆ decline in population/infrastructure ratios;
◆ wilderness experience declining;
◆ grandeur of Falls reducing.

Main beneficial environmental trends:

◆ increasing job opportunities;
◆ increasing per capita incomes;
◆ increased foreign exchange generation;
◆ increased environmental awareness and cross-border co-operation especially regarding resource management.

Main causes of environmental impacts:

◆   continuing increase in number of tourists and infrastructure requirements;
◆   revenue leakage and lack of re-investment in infrastructure;
◆   lack of local authority capacity to control/manage development, in-migration and environmental changes;
◆   lack of local, informal initiatives.

Specific causes of individual impacts:

◆   increased aerial flights (fixed wing aircraft, helicopters and micro-lights);
◆   too many boats on the Zambezi river and attendant jetties, etc.;
◆   growth in agriculture;
◆   use of the Zambezi for water abstraction and hydro-power.

Likely future significant environmental impacts (direct tourism-related projects):

◆   continuing and substantial increase in tourist numbers with a significant growing proportion of low-budget 'back-packing' visitors;
◆   increase in 'low-impact' lodges and hotels;
◆   increase in number of international class hotels/lodges and extension of existing facilities;
◆   construction of more 'cultural' villages.

Likely future significant environmental impacts (other projects):

◆   airport upgrading/expansion at Livingstone and Victoria Falls;
◆   road upgrading (e.g. Lusaka to Livingstone; Zambia to Angola/Botswana);
◆   urban expansion at Victoria Falls and Livingstone;
◆   enhanced hydro-power generation and water abstraction (with possible pipeline to Bulawayo);
◆   increased electrification (Zambian side);
◆   one or two bridges;
◆   abattoir development.

Following the initial workshop an open, participative approach to stakeholder analysis was designed and implemented to establish all concerns, problems and management measures. This included interviews with community leaders and members, community organisations, business leaders and staff of local agencies (Meynell 1997). Public consultative meetings were held and opportunities for private interviews were made available. A final consultative meeting was held to present findings and allow for agreement on recommendations and management measures.

Analysis of the issues identified in the scoping workshop was conducted by formulating a series of problem trees, an approach commonly used in development appraisal (Deutsche Gesellschaft für Technische Zusammenarbeit (GTZ) GmbH

TABLE 8.2   Scenarios for annual visitor numbers to the Victoria Falls area

Scenario	Zimbabwe	Zambia	Total
Present	220,000	66,000	286,000
Low growth	440,000	119,000	559,000
Medium growth	660,000	165,000	825,000
High growth	1,100,000	300,000	1,400,000
Super growth	1,100,000	760,000	1,860,000

Source: Meynell (1997)

1988). The analysis indicated that change in visitor numbers was the key criterion influencing the quality of environment, urban and rural life and the tourism 'experience' and so development scenarios were set using increased numbers of visitors, extrapolated from present figures (Table 8.2). From these scenarios it was possible quantitatively to predict changes in demand for: services such as schools, housing, water and sewage and solid waste treatment; tourist facilities such as new hotels and lodges, river cruising, rafting; and utilisation of resources (agricultural land and forests for fuel wood and curio carving) (Meynell *et al.* 1996). This in turn led to the identification of a series of impacts and their assessment according to whether they would be beneficial or adverse. Impacts were described as low, significant or major and presented as a series of matrices. Cumulative impacts for each scenario were calculated by summing the numbers of impacts for each activity/issue.

The analysis concluded that the overall limit to tourism-led growth for the Victoria Falls area was somewhere between the low- and medium-growth scenarios. In evaluating its approach, the study team considered that good guidelines were created by setting time and geographical limits at an early stage and by the insistence on carefully planning and implementing a participative stakeholder analysis. At the end of the study, in March 1996, the governments of Zambia and Zimbabwe committed themselves to establishing a cross-border institution for co-ordinated environmental planning and monitoring.

Questions for thought

1. What are the limitations of EA at the project level compared to SEA?
2. How important is cumulative effective assessment to SEA?
3. Discuss the merits of the application of SEA to a road development scheme.
4. What are the problems of conducting SEA across local, regional or national political borders?
5. What would be your criteria for defining the limits of an SEA study?

EA in practice

◆ **Introduction**   172
◆ **Example 1: Chipboard manufacturing plant**   172
◆ **Example 2: Pig breeding centre**   184

Introduction

The two EA case studies described in this chapter deal with a proposed wood chipboard plant in Ayrshire, Scotland, and a pig breeding centre in Inverness, Scotland. The examples delineate the study methodology and assessment findings.

EXAMPLE 1: CHIPBOARD MANUFACTURING PLANT

Eltimate Ltd (a subsidiary of Egger (UK) Holdings Ltd) submitted a planning application to Cunninghame District Council (CDC), Ayrshire, for permission to develop a chipboard manufacturing plant at South Gailes, near Irvine. Due to the nature of the proposed development an EIA was undertaken in accordance with the Environmental Assessment (Scotland) Regulations 1988 to support the planning application (Aspinwall & Company 1995). Each element of the proposed development which specifically relates to its construction and operation was examined to determine the environmental effects and, where necessary, suitable mitigation measures. In addition to specialist staff employed by Aspinwall & Company independent studies were undertaken by Natural Resource Consultancy (NRC) – ecology and nature conservation and The MVA Consultancy – traffic and access.

Purpose of the development

The purpose of the development was to manufacture quality chipboard products on a twenty-four hour continuous basis throughout the year. The plant was expected directly to employ in the order of 120 people, with an annual chipboard production of 400,000 m^3 (approximately 270,000 tonnes) in board widths up to a maximum of 2.2 m.

Site context

The site of the proposed development was at South Gailes, at the southern extremity of Irvine New Town (Figure 9.1). The area of South Gailes was formerly identified as a development area with potential to accommodate major population growth, but it was re-designated on the basis that land better suited to public and private housing development was identified elsewhere in the New Town. The establishment of a paper mill in the late 1980s adjacent to the development site reduced the attractiveness of the area for such development. Accordingly, the 1987 Irvine New Town Master Plan

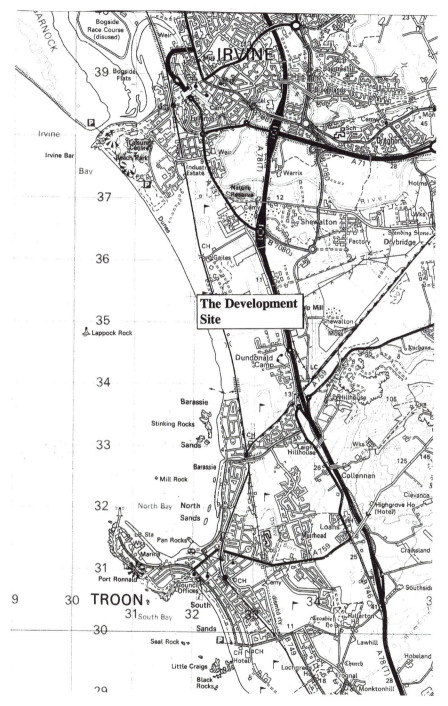

FIGURE 9.1 Site location map

Review recommended that alternative open space proposals, in addition to existing forestry and marginal agricultural use, should be considered. In recognition of this review, CDC zoned South Gailes as 'Countryside' in its adopted Irvine Kilwinning Local Plan of 1989.

The application site extended approximately 21 ha in area and was under let for agricultural grazing on a short-term lease. The majority of the South Gailes area could be described as open space, woodland and marginal agricultural land. The site is adjacent to the Firth of Clyde, and is low lying with a gently declining gradient westward. Towards the south is low-density housing and recreational land use, while to the north is a designated conservation area, woodland and a golf course. The area to the south of the proposed development was used as a military camp and training base during the First and Second World Wars. To the south of the site is a zone of residential buildings. There are also three golf courses in close proximity to the site.

The site area itself constitutes an area of flat and largely featureless landscape where marsh and peatland vegetation has been mostly degraded due to agricultural grazing. There are areas of woodland which are in close proximity to the proposed site, all of which were established as a result of land reclamation/rehabilitation work. Immediately to the east of the site is a strip of mixed woodland planting consisting of trees planted between 10 and 15 years ago. Within a nearby golf course is a Site of Special Scientific Interest (SSSI) which was designated to protect the characteristic sand dune ecology which includes a rich variety of invertebrates. The Gailes Marsh Natural Heritage site lies adjacent to the north-western boundary of the site and includes an area of established woodland, grassland, and marshland important for wading birds and wildfowl.

Legislative requirements

Under the Environmental Assessment (Scotland) Regulations 1988, certain types of projects require the developer to accompany a planning application with an ES, comprising environmental information which the planning authority is bound to take into consideration in its determination of the planning application. Schedules 1 and 2 of the Regulations list the types of project which are subject to an EA. Manufacture of chipboard falls within the scope of Category 8(b) of Schedule 2, i.e. '*the manufacture of fibre board, particle board or plywood*', where an EA is not mandatory, but is dependent on the likelihood of 'significant environmental effects'. Guidance has been issued by the Scottish Development Department on the types of development which are likely to require EA, and indicative criteria for Schedule 2 projects. For manufacturing industry, new sites requiring in the range 20 – 30 ha of land for development may well require EA, in addition to those expected to discharge waste and emit pollutants. The requirement for an EA is also dependent on the location, nature and significance of the emissions. The CDC requested that an ES accompany the planning application.

The following principal development plans which contained guidance relevant to the proposed development were reviewed with respect to the proposal: Strathclyde Structure Plan – The Consolidated Structure Plan, Corrected Edition 1991; the Irvine/Kilwinning Local Plan, adopted August 1989; the Irvine Kilwinning Local Plan. Alteration Nos 1 and 2 to the Local Plan; and the South Gailes Submission under Section 6 (1) of the New Towns (Scotland) Act 1968, Irvine Development Corporation (IDC), February 1995.

The specific locational requirements for the proposed chipboard manufacturing plant were: a large, open and flat area for construction of production facilities and ancillary building and storage areas; the need for sufficient buffering distance from potentially sensitive receptors (e.g. residential areas) to prevent an industrial/residential interface, potential public health impacts and environmental nuisance; a viable rail linkage to the plant to allow the export of finished products to the main centre of furniture manufacture concentrated around the M62 corridor stretching from Hull to Liverpool, and for possible European export; the need for suitable access to utilities for optimal operational efficiency during production, including power, communications, water supply and sewerage; suitable road infra-structure and access arrangements for supply of local raw materials (roundwood and sawmill residues), export of finished products and movement of staff; the need for a suitably skilled local workforce and sufficient area for potential future expansion.

Within the context of these specific locational requirements a site selection procedure was initiated by Eltimate Ltd in conjunction with Locate in Scotland at national, regional and local levels. Following the study the South Gailes site was considered to be compatible with the specific locational requirements because it was considered large enough to accommodate the plant and potential future expansions required to support product upgrading (with associated employment benefits); it was not constrained by adjacent developments; the area was also considered to be of a suitable gradient for construction requirements; the development of the plant in this location was considered to be compatible with the presence of an adjacent paper mill which already detracted from the rural and local amenity value of the area; the site was not located immediately adjacent to a high density residential area which thereby limited the potential for significant public health impacts and levels of environmental nuisance – outwith the regulatory operating constraints and emission controls incorporated into the plant design. The prevailing wind direction, from the south west, was also regarded as being beneficial with respect to any potential airborne emissions on nearby housing to the south and south west of the site; no significant constraints were envisaged with regard to the provision of utilities, including power, communications, water and sewerage; and the site was well connected to the local and regional road infrastructure and a railway line. In addition, it was considered that the site's proximity to the port of Troon would provide the opportunity to import timber from the Western Isles of Scotland.

The development

The plant would utilise the latest continuous press technology, where the design incorporates the appropriate pollution control equipment to limit atmospheric emission levels to within the authorisation conditions set out in the Secretary of State's Guidance Note PG 6/4 (94). Chipboard is manufactured from thin dried flakes of wood blended with synthetic formaldehyde resins and compressed under heat to form flat panels. The six defined stages within the manufacturing process are wood particle preparation and storage; particle drying; particle separation and grading; resin application; hot pressing; and calibration and finishing.

The intended construction period for the proposed chipboard plant would be of fourteen months' duration. During this period site preparation would be carried out involving installation of drainage, service infrastructure and building foundations. The construction phase would thus involve the creation of access and site traffic routes and movement of plant and equipment onto the site as well as routine construction operations. A site boundary fence would also be erected. External works would involve the creation of hard standing areas for the log park, as well as car parks and vehicle circulation areas. It would also involve the establishment of the landscape works proposed for the site, which included the creation of a minimum 20 m planted area along the south and north site boundaries, as well as some internal planting around car and lorry parks. It was proposed that native hardy stock requiring minimal maintenance be planted.

Consultation and scoping of environmental issues

Baseline studies

Baseline monitoring studies were conducted to supplement available information relating to the proposed development site. These studies included a landscape and visual survey; ecological survey; traffic survey; ambient noise monitoring and air quality monitoring.

Environmental scoping

The scoping study consisted of a review of development proposals and consultations with statutory, and other, authorities. In addition, current government guidance and a number of strategic planning policy documents produced by the Regional and District Councils were consulted and the key environmental issues identified. Consultations were undertaken with a number of different bodies to establish their concerns, to obtain baseline environmental information, and to identify the potentially significant issues associated with the proposed development. The statutory consultees involved included: Strathclyde Regional Council, Planning Department; Kyle and Carrick District Council, Planning Department; Health & Safety Executive; Scottish Natural

Heritage (SNH); Clyde River Purification Board; Irvine Development Corporation; Secretary of State for Scotland (Scottish Office: Her Majesty's Industrial Pollution Inspectorate (HMIPI); Scottish Office: Development Department; Scottish Office: Agriculture, Environment and Fisheries Department; and Historic Scotland); Strathclyde Regional Council, Strathclyde Water Services; Strathclyde Regional Council, Roads Department, Planning Department, Environmental Health Department; Scottish Wildlife Trust; Scottish Ornithological Club; Royal Society for the Protection of Birds (RSPB); and The Coal Organisation.

The environmental issues raised during the scoping exercise and associated consultations were addressed in the environmental statement and the accompanying appendices.

Assessment of impacts

Landscape and visual assessment

The study assessed the potential effects on landscape and visual amenity resulting from the proposed development. Both positive and negative effects of the proposal were examined during the construction and operation of the proposed plant.

The study established the baseline conditions in terms of the existing landscape and visual context of the site; defined the likely sources of landscape and visual effects arising from the proposed development; analysed the visibility of the proposals from viewpoints in the vicinity of the site; assessed the significance of the potential landscape and visual effects arising; identified mitigation measures which were or could be incorporated into the scheme and assessed any residual landscape and visual consequences.

The existing landscape context was described in terms of relevant planning policies as well as a classification of the landscape according to main landscape character zones, topography, land use and settlement and communications in the area surrounding the site. A detailed description was also given of the existing landscape of the site. The proposed landscape scheme was described and the potential effects of the proposal on the landscape of the site surrounds were evaluated.

A zone of visual influence (ZVI) was defined as the area around the site within which the proposed development would be a significant visual feature and component of the overall view. Within the defined ZVI, sensitive locations were identified and the view from these locations analysed and assessed for potential visual effects. Three categories of sensitive viewpoint were identified: residential buildings; recreational land use; and the road and rail network.

Environmental consequences

The landscape assessment demonstrated that both the construction and operation stages of the proposed development would constitute a significant change in the character of the existing landscape in the immediate vicinity of the application site, and in particular of the landscape on the west side of the A78(T). When considered in the wider context of the landscape of the surrounding area, the proposed development was less intrusive. There were a number of industrial developments which occurred in the predominantly rural landscape between the urban areas to the north, east and south of the application site, giving rise to a varied landscape character in the area.

During the construction stage the visual effects of the construction stage of the proposed development would vary from *severe* to *insignificant* when considered from the sensitive viewpoints which were analysed. Most severely affected would be the residential properties immediately to the south of the site. The effect of the construction stage on recreational areas was less significant, largely due to the distance of these areas from the application site and intervening landforms or woodland which would screen most of the construction activity. The construction of the proposed railway link would, however, have a *moderately negative* effect on views from the northern part of the Kilmarnock (Barassie) Golf Course (see Figure 9.1). From the road and rail network, a *moderately negative* visual effect would arise during the construction stage over part of the A78(T) to the south of the application site, resulting from the change in character of the view westwards from this location. Views from the footbridge over the A78(T) adjacent to the northern boundary of the application site were also considered to be *substantially negatively* affected. The effect on all other parts of the road and rail network analysed varied from *slight* to *insignificant*.

During the operational stage of the plant the visual effects varied from *severely negative* to *insignificant*. Once again, the most severely affected viewpoints were those associated with the residential areas to the south of the application site due to the proximity of the proposed industrial development and the change in views occurring as a consequence. Views from the recreational areas considered were mostly considered to be *slightly or moderately negatively* affected by the proposed development. The northern part of the Kilmarnock (Barassie) Golf Course would be in close proximity to the proposed railway link which would cut across agricultural land currently dominating views from the northern part of the course. The most sensitive viewpoint on the road and rail network in the vicinity of the application site was on the A78(T) to the south of the proposed development, which was considered to be *substantially negatively* affected, due to the proximity of the proposed development and the change in character which it would bring about in the overall view westwards from the northbound carriageway of this section of the road. The footbridge over the A78(T) adjacent to the application site was also considered to be *substantially negatively* affected by the operational stage of the proposed development for the same reasons. The visual effect on views from all other road and rail locations considered in the assessment was either *slightly negative* or *insignificant*.

Ecological assessment

NRC undertook an ecological assessment of the application site. Consideration was also given to potential impacts on adjacent habitats and designated conservation areas.

The Irvine Development Corporation (IDC), The Scottish Ornithological Club, The Scottish Wildlife Trust (SWT) and Scottish Natural Heritage (SNH) were approached by NRC for existing survey information. No other significant additional sources of data were anticipated. A field survey of the application site and surrounding area took place during late September/early October 1995. The application site was surveyed in more detail than the surrounding area. A Phase 1 Habitat Survey of the site and its surrounds was carried out in accordance with the Nature Conservancy Council (NCC) Phase 1 Handbook. Particular attention was given to the condition of watercourses and wetlands. Quadrat information was gathered on the development site in order to classify it according to the National Vegetation Classification (NVC). As full a species list as possible was made of the plants within the application site, although the survey took place late in the year. A local bird expert was consulted regarding ornithological interest associated with the site: of particular interest was the Golden Plover (*Pluvialis apricaria*).

Environmental consequences

The primary impact on the ecology of the South Gailes area was associated with land take for the construction of the chipboard plant within the application site. Due to the bird life (particularly the Golden Plover) and botanical interest associated with the site, impacts, in the absence of recommended mitigation measures, were considered to be potentially significant, negative and long term. Effects on the surrounding area would also be anticipated, as the habitat within the application site formed part of the more extensive dune system in the local area. However, there was a potential to compensate for any negative effects by implementing the mitigation measures.

Mitigation measures were recommended to reduce the degree of impact on ecology during the construction and operational phases of development. In order to offset the impacts associated with the development it was recommended that the use of adjacent land to the west of the application site be considered for conservation purposes. Following the findings of an initial impact assessment by NRC, Aspinwall & Company met with the Royal Society for the Protection of Birds to discuss potential impacts and potential mitigation measures. Based on these discussions and further available information, the assessment was updated, in conjunction with NRC, to agree suitable mitigation measures. Eltimate Ltd indicated its general support for adjacent habitat being used for conservation purposes. With the implementation of a suitable conservation plan and grazing regime, it was possible that the ecological value of remaining adjacent habitats could be enhanced.

Traffic assessment

The assessment dealt with road and traffic issues arising in connection with the proposed development and included an appraisal of existing conditions; the derivation of traffic generation predicted as a result of development; an assessment of the impact

TABLE 9.1 Summary of peak hour vehicle movements

Source of generation	Arrivals	Departures
Employees	30	30
Raw materials in (max.)	7	7
Product out (max.)	5	5
Ancillary activities, say	3	3
Total	45	45

resulting from the addition of generated traffic to existing traffic and a review of requirements for measures to accommodate generated traffic. The assessment was carried out following discussion with officials from Strathclyde Regional Council (SRC) to establish the scope and detail of the work. A preliminary estimate of traffic generation was prepared and submitted to SRC so that the scale of traffic impact could be appreciated. It was agreed that the key requirement would be to assess firstly the geometry and capacity of the site access points and secondly the capacity and geometry of the old Ayr Road arm of the Meadowhead Roundabout (see Figure 9.1) to accommodate use by large articulated vehicles. The impact on traffic conditions and the local road network in the South Gailes area was assessed in relation to plant operation only, precise construction details with regard to movement of vehicles being unavailable. However, recommendations were provided to alleviate impacts on local traffic as a consequence of construction plant access to the site. A summary of peak hour vehicle movement activities is given in Table 9.1.

At the site accesses there would be a maximum total of forty-five arrivals and forty-five departures split between two accesses in any hour. Experience dictated that there would be no capacity problem in accepting such flows on a road carrying 400–500 vehicles.

Environmental consequences

The conclusions of the traffic assessment for the proposed development could be summarised as follows: the plant location would bring an increase in traffic to the local road network; this increase is very modest when compared to existing traffic on the network; traffic generated by the development could be accommodated without difficulty on old Ayr Road and at Meadowhead Roundabout (see Figure 9.1); it would not be necessary to carry out any works on Meadowhead Roundabout to accommodate development traffic; and shift working at the proposed plant would avoid any significant additional traffic impinging on the existing network peak period. Overall, it was concluded that traffic generated in association with the construction and operation of the plant would result in insignificant or minor impacts on existing traffic conditions in the South Gailes area. The use of a railway connection would further reduce potential road traffic impacts.

Noise from the proposed development had the potential to cause adverse impacts on the local community and environs. Noise from the plant was examined in relation to the possible impacts at locations identified as being potentially sensitive to noise. This allowed the noise impacts to be assessed by comparing noise levels due to the proposed activity with relevant guidance and also with existing noise levels. The acceptability of such impacts was assessed and, where necessary, suitable mitigation measures proposed.

All road traffic associated with the construction and operation of the proposed development would enter and leave the site via the old Ayr Road and would use the major road network, including the Irvine Bypass A78(T), to approach and leave the area. Other local minor roads would not be used by heavy site traffic. Although there would be additional traffic on the section of the A78(T) directly adjacent to the site, there would be no additional noise impact for the properties closest to the old Ayr Road in this area. This was because the existing volume of traffic on the A78(T) would remain dominant, and any increases on the old Ayr Road would not be significant in terms of generated noise levels. The traffic generated by the operation of the site amounted to a total of only sixty light vehicles and thirty heavy vehicles per (peak) hour, while the existing flows were many times this volume. An increase in flows of 25 per cent would increase noise levels by only 1 dB(A), this being the minimum noise increase likely to be perceived. As long as site traffic was not routed on any other local roads, it would not result in a measurable change in flows on the major road network. Therefore, the detailed assessment of noise impacts generated from site related traffic was not considered necessary.

The construction phase of the development would take approximately fourteen months. During this time the whole site would, at various times, be subject to construction activities. Typical construction activities would involve the use of: excavators; lorries; dump trucks; mobile cranes and generators/lighting equipment. It was also anticipated that piling would be necessary in the construction of the site foundations, although the extent and duration of piling activity was not established. The standard method for predicting noise impacts associated with open construction sites is BS5228: Part 1: 1984: 'Noise control construction and open sites'. Part 1 of this British Standard, 'Code of practice for basic information and procedures for noise control', provides a method for combining the contribution of noise from a number of individual items of plant, taking into account their locations, their sound power levels, and the percentage of the time that they are operating, or percentage 'on-time'.

With regard to the operational phase the proposed development would include a number of items of fixed plant which would generate noise at sufficiently high levels to warrant further assessment. Details of noise-generating items of plant to be introduced on to the site were provided by the developer, as was information on the noise levels prevailing at a specified distance from a similar plant monitored elsewhere. This information was used to predict noise levels at noise sensitive receptors (NSRs) near to the proposed development. Potential noise impacts arising from fixed items of plant were assessed using the methodology recommended by BS4142: 1990. This methodology involved the comparison of predicted operational noise levels with existing noise levels at the facade of identified sensitive receptors. Assessments carried out in this way made it possible to predict the likelihood of complaints arising from operational noise. The noise contribution from mobile noise sources on the site was added to that due to fixed plant, and the combined impact assessed. The Environmental Health Department of CDC indicated that they would require operational noise from the proposed development to be controlled sufficiently to result in no net increase over the existing ambient noise levels at the sensitive receptors concerned. In order for this requirement to be achieved, noise levels at these sensitive receptors due to the operation of the proposed development alone would need to be 10 dB(A) lower than the monitored baseline. These figures were considered to be unreasonably and unnecessarily low, as the sensitive receptors concerned would be adequately protected by noise limits which are higher. It was therefore considered reasonable to adopt noise standards which are consistent with BS4142: 1990. BS4142: 1990 states that noise from a proposed development which is 5 dB(A) higher than the existing background noise level is of marginal significance. A level which is, say, 3 dB(A) above the background would therefore offer a reasonable level of protection to local residential properties. This approach has been used elsewhere to allow a reasonable planning standard to be achieved.

Environmental consequences

Noise arising from the construction of the proposed development could be controlled by a planning Section 60/61 Agreement, and additionally mitigation measures would assist in minimising undesirable noise impacts at the noise-sensitive receptors to the south of the site. With regard to the operational phase it was shown that if proposed mitigation measures were provided, the specific noise sensitive receptors identified would not be exposed to significant noise impacts resulting from the operation of the proposed development.

Air quality assessment

The impact of emissions to atmosphere from the proposed plant's operation and potential fugitive emissions from construction activities was assessed. Where necessary, mitigation and monitoring protocols were also detailed in order to reduce air quality impacts to acceptable and agreed levels. The study firstly assessed the air quality standards relevant to the study, followed by the air quality for the locality and finally predicted the impact of emissions to air using atmospheric dispersion modelling techniques. Table 9.2 compares predicted levels, where possible, with recognised air quality criteria standards.

TABLE 9.2 Comparison of predicted ground-level concentrations from the proposed plant with recommended air-quality criteria (μgm^{-3})

	1 hour	24 hour	Annual
Particulates			
predicted level	3.65 – 152.92	0.62 – 26.0	0.14 – 5.70
baseline level		34.0	21.2
standard	—	150	50
VOCs			
predicted level	19.08 – 30.94	3.24 – 5.26	0.47 – 0.76
baseline level		<5	<5
standard	—	—	—
Formaldehyde			
predicted level	2.39 – 4.77	0.41 – 0.81	0.06 – 0.12
baseline level		<5	<5
standard	100	—	25
Total aldehydes			
predicted level	2.39 – 4.77	0.41 – 0.81	0.06 – 0.12
baseline level		<2	<2
standard	—	—	—

Environmental consequences

The study showed that air quality for the Irvine and South Gailes area was within recommended air quality criteria. The addition of emissions to air from the proposed plant was unlikely to cause these criteria to be exceeded. Predicted ground level concentrations were unlikely to be detrimental to human health.

Surface and groundwater assessment

The study examined the potential for the chipboard manufacturing plant to result in contamination of surface and groundwaters in the vicinity of the site and, where necessary, recommendations for suitable mitigation measures to reduce impacts to acceptable levels were made. A review of surface and ground water features within the vicinity was undertaken, together with an assessment of on-site drainage requirements and local facilities for the management and disposal of aqueous emissions associated with site operations.

Environmental consequences

Provided that suitable mitigation measures were implemented during the construction phase of the plant, significant or long-term impacts on surface or ground waters were not envisaged. It was recommended that the construction contractor liaise with the regulatory authority and implement the control measures produced in the relevant pollution control guidelines for civil engineering works to avoid any unacceptable impacts. The design of the drainage system would ensure the segregation of clean and dirty waters generated during the operation of the plant, which together with additional mitigation measures specified should ensure that no unacceptable impacts on surface and ground waters arise during the operation of the plant. No constraints with regard to the operational capacity or related discharge consent were anticipated.

Regular monitoring during the construction and operational phases of development was recommended to ensure that any deterioration of a nearby stream was detected and remediated in conjunction with the regulatory authority at the earliest opportunity.

Conclusion

The developer submitted the EA with its planning application following completion of the studies. It subsequently withdrew the application following opposition to the development from IDC and local residential groups. The developer later gained planning permission for the project at another site nearby at Barony in East Ayrshire, Scotland. The format and the content of the revised ES were based on the original study for South Gailes.

EXAMPLE 2: PIG BREEDING CENTRE

PIC UK were seeking planning consent to develop a pig breeding centre near Inverness, Scotland. Aspinwall & Company was commissioned by PIC to undertake an independent EA (Aspinwall & Company 1996b) to accompany the planning application for the new development to be submitted to Highland Regional Council (HRC). The objectives of the EA were to identify the potential environmental effects of the proposed development, taking into account the characteristics of the development and the local environment, and the views of the local authorities and consultees with responsibilities for the environment; to interpret the nature and extent of potential impacts and to present the findings of the assessment in a formal statement which is available to the statutory authorities and the public.

Purpose of the development

The purpose of the development was to house a herd of 2,400 breeding sows with an accompanying eighty boars which would be used as stock for artificial insemination. In total, including progeny, it was anticipated that there would be approximately 24,000 pigs present at the centre at any one time. The centre would directly employ up to thirty-five persons, with employment for a further twenty-five generated indirectly from activities associated with slurry removal, feed manufacture, farm maintenance and provision of support services. There would also be between twenty and thirty jobs generated by the construction of the buildings over a twelve-month period.

Site context

The site of the proposed development was located approximately 6.5 km south of Inverness City centre. The 950 acre site encompassed Tom Fat Woodland, Nairnside and Inverarnie plantations on the upland of Drummossie Muir (Figure 9.2). The site is a rural location and was afforested with predominantly coniferous tree species. Evidence existed of agricultural activity prior to afforestation. The surrounding area was also dominated by similar coniferous plantations, together with livestock grazing (including sheep and cattle). No significant industrial activities existed in the vicinity of the site, although two sand and gravel quarries were located to the south-east and east of the site at a distance of less than 1.5 km.

There were a number of residential properties in the vicinity of the application site, which included approximately twelve individual residential properties, smallholdings and farms in the vicinity of the south-eastern boundary along the Daviot Road heading north-east from the junction with the B861 at Balnafoich towards the Mains of Faillie; approximately ten individual residential properties, smallholdings and farms adjacent to the western boundary of the application site along the B861. Some properties were recently constructed, or were currently under construction; and isolated properties to the north of the application site, within a distance of 2 km, included Balvonie of Leys, Newton of Leys, The Grange and Leys Castle. Milton of Leys, approximately 2 km north-east of the application boundary, had approval for further significant residential development, although no development schedule was known to exist at the time of the study.

The majority of the site consisted of mature coniferous tree species with intermittent open areas consisting of heather moorland. The area around Loch Caulan, beyond the application site boundary, comprised upland bog vegetation indicative of a high water table. The application site did not encompass any areas designated for their environmental quality and sensitivity.

Legislative requirements

In accordance with the Town and Country Planning (General Development Procedure) (Scotland) Order 1992, the planning application would be submitted to HRC Planning Department. Under the Environmental Assessment (Scotland) Regulations 1988, certain types of projects required the developer to accompany a planning application with an ES, comprising environmental information which the planning authority is bound to take into consideration in its determination of the planning application. Schedules 1 and 2 of the Regulations list the types of project which are subject to EA. Breeding of pigs falls within the scope of Schedule 2, being

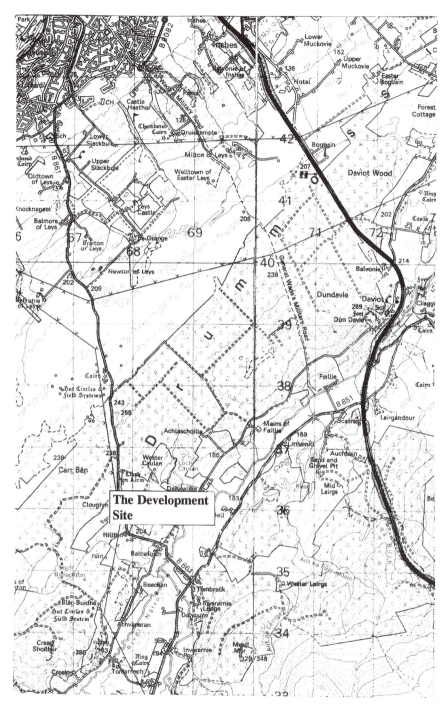

FIGURE 9.2 Site location map

defined as '*pig-rearing*'. Scottish Development Department (SDD) Circular 13/1988 provides an indication of the scale of development likely to require an EA which, in the case of the application site, includes intensive pig rearing installations housing more than 400 sows and 5,000 fattening pigs.

The Control of Pollution Act (COPA) 1974 had relevance to the operational phase of the proposed development with regard to protection of water and land resources and environmental nuisance.

Further statutory duties with regard to the quality of drinking water supplies were the responsibility of the water authorities under the Water Act 1989. Environmental Health Authorities had a duty to be informed of the quality of public and private water supplies with regulations on the monitoring, sampling and analysis of private supplies. The relevant authority in relation to the application site was Inverness District Council (IDC).

The Buildings Standards (Scotland) Regulations 1990 included requirements for dungsteads and farm effluent tanks (including slurry and silage effluent tanks). These requirements included measures intended to prevent contamination by the contents.

Odours arising from livestock housing and the storage and spreading of manures and slurries had the potential to result in environmental nuisance at nearby residential properties. Such nuisance was controlled by the Public Health (Scotland) Act 1897, and local authorities were obliged to inspect their areas to detect any statutory nuisances and to take practicable steps to investigate complaints of statutory nuisance which are made to them.

Planning policy context

The principal development plans which contained guidance relevant to the proposed development included the Highland Region Structure Plan 1990; also Alteration No. 1 (Forestry) 1994 and the Strathdearn, Strathnairn and Loch Ness East Local Plan, Final Draft October 1994.

The Structure Plan was a general statement of strategic planning intentions for the Highlands. It provided development guidance for the public and private sectors, highlighting areas where opportunities exist and where steps should be taken to safeguard and enhance the rural environment. It provided a framework for local plans with respect to development control issues and was a material consideration in the determination of planning applications. Policies contained within the Structure Plan were reviewed in relation to agriculture, forestry and the environment. There were no specific Policies, Recommendations or Strategies which precluded the development of a pig breeding centre or livestock installations in general.

The Local Plan, consisting of written policies and specific land-use allocations, was more detailed than the Structure Plan. The Strathdearn, Strathnairn and Loch Ness East Local Plan was in Final Draft, dated October 1994. The application site was located within a restricted development zone, where the surrounding area was regarded as pressurised countryside, mainly as a result of residential development associated with Inverness.

The economic benefits of the development were in general accord with regional and local policy in terms of enhancement of the local economy and alleviation of unemployment. The area was within a restricted development area, mainly with regard to suburban development associated with Inverness. The area was also recognised on the basis of its local amenity and conservation value, although the coniferous plantations associated with Drummossie Muir could be regarded as being of marginal value in this context. It was acknowledged in the Local Plan that HRC would encourage diversification, innovative land management and development schemes embracing specialised farming, subject to adequate access and compatibility with neighbouring uses and amenity. It was considered, therefore, that the proposed development was generally compatible with local planning policy, with the proviso that environmental impacts associated with its construction and operation were maintained within acceptable levels.

The requirements for the pig breeding centre were: a large, open and flat area for construction of pig breeding facilities and ancillary building and storage areas; a sufficiently remote and secure site to minimise the risk of intrusion and incidence of pig infection; sufficient buffering distance from potentially sensitive receptors (e.g. residential areas) to prevent a residential interface and alleviate potential public health impacts and environmental nuisance; suitable access to utilities for optimal efficiency during operation, including power, communications and water supply; suitable road infrastructure and access arrangements for supply of feed materials, export of livestock and movement of staff and a suitably skilled local workforce.

It was within the context of these specific locational requirements that a site selection procedure was initiated by PIC at the international and national level. Within Scotland four sites were initially identified and assessed based on criteria contained within Schedule 3 of the Environmental Assessment (Scotland) Regulations 1988. The potential environmental constraints of each site and their potential significance were assessed. From this it was apparent that environmental constraints to development existed for each of the sites. However, on balance, it was considered that the constraints relating to the alternative sites were potentially more significant than the site at Drummossie Muir. The site was considered to be compatible with the specific locational requirements because it was considered large enough to accommodate the main pig breeding buildings and associated facilities; the development of the centre was considered to be broadly compatible with the local amenity and conservation value of surrounding land uses as it would be incorporated within existing mature coniferous plantation, thereby screening the development from the surrounding area; the site was sufficiently remote and secure to minimise the risk of intrusion and the incidence of pig infection and the site was not located immediately adjacent to a high density residential area which thereby limits the potential for significant public health impacts and levels of environmental nuisance. The prevailing wind direction, from the southwest, was also regarded as being beneficial with respect to any potential odour emissions on nearby housing to the south and south-west of the site. No significant constraints were anticipated with regard to the provision of utilities, including power, communications and water and no overriding constraint was envisaged with regard to management of liquid effluents and sewage. The site was reasonably connected to the local and regional road infrastructure.

The development

The site would house 2,400 breeding sows with an accompanying eighty boars which would be used as stock for artificial insemination. In total, including progeny, it was anticipated that there would be approximately 24,000 pigs present at the farm at any one time. Annual production of animals would be 12,000 pigs for breeding purposes and 36,000 for meat production. The inputs and outputs of materials and livestock are summarised in Table 9.3.

TABLE 9.3   Inputs and outputs of livestock and materials

Item	Quantity	Item	Quantity
Inputs		*Outputs*	
Feed	5,000 tonnes/year	Solid muck	8,000 cubic metres/year
Straw	3,000 tonnes/year	Slurry	35,000 cubic metres/year
Water	144,300 litres/day	Finished pigs	36,000/year
Electricity	6,500 units/day	Breeding pigs	12,000/year
Labour	35 people		

The building layout would comprise three discrete sites (i.e. pig breeding, pig rearing and accommodation). The total area of these three sites was approximately 150,000 square metres and the total roofing area was about 35,000 square metres.

Road access to the site for emergency vehicles and staff would be via a junction off the B861. Service and animal transportation vehicles would access the site via the Daviot Road south of the site linking the B861 to the A9.

Consultation and scoping of environmental issues

The study consisted of a review of development proposals and consultations with a number of relevant authorities. In addition, current government guidance and a number of strategic planning policy documents produced by the Regional and District Councils were consulted and the key environmental issues identified. Authorities consulted included: IDC Environmental Health Department; HRC Highways Department; HRC Planning Department; HRC Department of Water and Sewerage; HRC Regional Archaeological Unit; Highland River Purification Board (HRPB); Scottish Wildlife Trust; Historic Scotland; HMIPI; SNH and the RSPB.

Assessment of impacts

Landscape and visual amenity

The site was dominated by two high points to the north-east and north-west with a minor saddle creating a watershed between the River Nairn in the south and Allt Mor Burn, connecting to the Caledonian Canal in the north-west. The northern part of the area was the flattest with slopes approximately 1 in 50 while slopes approaching the Loch increase towards 1 in 12. The site had approximately 90 per cent cover of coniferous plantation with mainly single age species between 30 and 50 years old. The general height of the trees ranged from 10 m to approximately 20 m, with planting distances averaging about 4 m. Small clearings within the site were due to tree felling activities. Visual penetration at immediate ground level was approximately 50 m, assuming level ground. Within the application area the forest was dissected by a number of tracks planned on a grid iron system intersected by a number of fire breaks. Significant to the area were a number of archaeological features which include standing stones, chambered cairns, hut circles and the remains of field systems. These latter features were indistinguishable at ground level and were tree covered.

The site buildings made up three separate sites. These comprised sites 1 and 2, breeding and rearing facilities, and a staff hostel. Sites 1 and 2 had a minimum separation distance of 700 m, and separate vehicular access. Each group of buildings was located in the flattest areas of the site, taking advantage of existing access tracks where possible. Sites which cut across the contours were terraced to minimise earthwork requirements. Security to sites 1 and 2 was provided by 2 m high chainlink fencing. Lighting was proposed at low level to illuminate walkways and buildings with directional reflectors to minimise light 'spillage'.

Access to the individual sites was planned to take advantage of existing tracks together with the two main access points off the B861 and the Daviot Road. It was proposed that the existing tracks, to be used for access, would be given a surface of compacted gravel or quarry scrapings and would be widened locally to provide passing places for the larger delivery vehicles. The remaining gravel tracks and fire breaks would also be kept.

Overhead electricity poles were located to the south of the site. It was proposed that this electricity supply was extended onto the site to supply the development along the road edge. Water supply via the mains from Daviot would be extended underground to the site. It was proposed to make provision for these external services, together with internal services such as drainage, etc., in a common corridor which would follow improved roadways.

No specific public rights of way apparently existed across the application site, but parts of the site were used by locals for dog walking or rambling. The proposed pig breeding sites had individual security fencing to preserve bio-security and it was felt that public access would be impossible to control. To discourage general access over the whole site, to the public, and create a positive opportunity as part of this development, it was proposed that a footpath be provided, on a circular route, to link the archaeological sites, and provide views of, and safe access to, Loch Caulan.

The current coniferous forest was planted for commercial reasons only and contained few indigenous species of trees. As part of the overall development some of the coniferous stands would be removed to accommodate building and surface water holding ponds. To encourage more diversity, for wildlife, and amenity it was proposed, as part of a longer term plan, to introduce a more indigenous broadleaf structure to create a more natural woodland. The proposals were derived from the recent National Vegetation Classification in Britain (NVC) and would follow the W17 classification (i.e. Upland Oak–Birch Woodland with Bilberry – Forestry Commission Bulletin 112 Creating New Nature Woodlands). The major recommended trees were Sessile Oak and Downy Birch with minor trees comprising Pendinculate Oak, Silver Birch, Holly and Rowan. Recommended shrubs included Hazel, Hawthorn and Juniper. Since one of the prime advantages of siting the development in the coniferous forest was visual seclusion, the development of the natural woodland should be undertaken during a 15-year woodland management period. This would have the following advantages: existing conifers would reach viable economic age; new tree planting would have time to reach a more mature height and effective screen before removal of coniferous trees; and financial costs are spread over a longer period giving the opportunity to reinvest returns from the commercial felling into this new planting and forest management.

The proposals for introducing natural woodland were restricted to boundaries, edges of proposed development, fringes of archaeological features and along access roads and footpaths. Within the compartments created by the new planting it was proposed that commercial forestry continue, to contribute to visual seclusion, especially during winter months and to maintain a revenue for reinvestment.

The visual assessment considered four prime points of view: the A9; the environs of Inverness; vicinity of houses along the B851 and B861; and potential points of public access within the forest. Within the visual range of the A9 the proposed site was predominantly hidden from view by either landforms or woodland planting adjacent to the roadside. It was considered that the impact on these views will be insignificant. The potential views from the environs of Inverness were diminished both by distance from the edge of Inverness (5 km) and by moorland in the middle ground. More important, visually, were the views from the B861 approaching the site from a distance of 2 km. The development site, i.e. coniferous forest, was seen as a strong line of trees near the ridgeline. It was considered that the impact on these views will be insignificant. The views of the development from the various houses were restricted by either landform, i.e. 250 m ridgeline shadow, or visual penetration limited by the density of forest planting. It was considered that the impact on these views would be insignificant. While the public would be discouraged from the forest areas containing the development sites, it was recognised that some filtered views from within the forest would be possible. Mitigation of this situation would include the use of screen banking and shrub planting using surplus overburden from the construction phases and the use of earth colours, e.g. dark green, brown or black for buildings' sides and roof, and feed silos. By the adoption of these principles the views of the development would be acceptable.

Environmental nuisance

The issues addressed included air quality including odour generation and ammonia emissions; noise and lighting. Where appropriate, issues related to the construction and operational phases of the farm development were addressed. Where significant impacts were identified suitable mitigation measures were, where possible, recommended to reduce residual impacts to acceptable levels.

Air

The principal air pollutants of concern in relation to pig rearing are odours and ammonia emissions. Agricultural activities which involve housed livestock and storing of wastes are generally those most likely to cause potential odour problems. Farming odours fall into two categories – those originating from processes such as grass and manure and the larger group of smells arising from animal houses and the biological breakdown of manures and other wastes. Animal houses which are force-ventilated can present problems in nearby communities but smells/odours arising from these are a matter of good house-keeping and management and, in general, do not present a significant problem. When manures are removed as a solid and stored in the open few odour problems arise until the store is broken up. An assessment of potential impacts on nearby sensitive receptors was undertaken on the basis of the location of potential emission sources, intervening topography and vegetation and prevailing wind direction.

The ventilation system of the pig houses would be natural/curtain sided and not forced in the service, gestation, acclimatisation, finishing, growing and test units. For the nursery units ventilation would be by cross flow means, and in the farrowing houses an extraction system would be used. The principal sources of odour and ammonia emission from the development would be the pig housing buildings and slurry/waste stores. The proposed development site would be situated within the woodland area of Drummossie Muir, on elevated ground at approximately 250 m above ordnance datum (AOD). The nearest housing to the site was situated to the east (Mains of Faillie), south-east (Achlaschoille) and south (Wester Caulan). Scattered homesteads lay further east and south of these premises. The proposed location of the pig farm would be approximately 50 m AOD relative to the nearest house. The nearest premises were about 600 m from the nearest potential odour and ammonia source. The premises would not be exposed for the majority of the time to potential sources of environmental nuisance as they were not in the pathway of the predominant wind direction. In addition, the premises would be screened by mature coniferous woodland, where the canopy of coniferous trees would provide partial attenuation to the dispersion of potential odours and ammonia emissions towards the areas.

Sensitive receptors which were less well screened by intervening coniferous plantation included Newton Leys and Fountainhead (north) and Balvonie Cottage and Balvonie of Leys (north-west). Potential new residential development at Milton of Leys at a distance of approximately 2 km north would also be potential sensitive receptors to nuisance. However, sensitive receptors to the north and north-west of the site experience prevailing winds for a limited period of time.

The nearest sensitive receptor situated within the direction of the prevailing wind direction of the site was approximately 3 km distant. All potentially exposed residences were more than 1.0 km from potential nuisance sources and at a lower height AOD. It was considered that due to the intervening distance from and topography of residential properties and the low occurrence of incidental wind direction, a significant environmental nuisance was unlikely to occur.

Noise

The construction phase of the development would take approximately 12 months. During this time several parts of the site would at various times be subject to construction activities. Typical construction activities would involve the use of bulldozers; diggers and excavators; and cement mixers. It was not anticipated that piling would be necessary in the construction of the site. It was also proposed that a Code of Construction Practice (COCP) be agreed with the local authority so as to control to acceptable levels noise emissions arising from construction activities.

All road traffic associated with the construction and operation of the proposed development would enter and leave the site via the existing minor road which runs from Daviot to Balnafoich and defined the southern boundary of the site. Other local minor roads would not be used by heavy site traffic, with the possible exception of the B861 leading to and from Inverness. Traffic generated by the construction phase would mainly be vehicles delivering materials. During the operational phase traffic generated would include deliveries of feed and straw, and removal of waste materials, and collection of pigs. These increases, while representing a change in overall traffic flows, were not considered to be substantial enough to warrant further detailed assessment. However, the timing of such vehicle movements was likely to be of more importance and required further consideration. Site traffic would not be routed on any other local roads, and it would not result in a measurable change in flows on the major road network. Therefore, it was considered that the assessment of noise impacts in these situations was not necessary.

With regard to its operation the proposed development would include a number of noise sources which required consideration. Noise monitoring was carried out at other farms operated by the developer in order to obtain relevant noise data. This data was used to predict noise levels at sensitive receptors near the proposed development. Additionally, noise data from BS5228: Part 1 was employed in the case of general heavy goods vehicle movements. Potential noise impacts arising from sources were assessed using the methodology recommended by BS4142: 1990. Baseline levels typical of a quiet rural location were adopted for the purposes of the BS4142 noise assessment. These values represent the lowest background noise levels that were likely to prevail in the vicinity of the development site, and were applied at each of the sensitive receptors. They were a daytime 30 dB $L_{A90(1-hour)}$, and night-time 25 dB $L_{A90(5-minute)}$.

It was established that noise from road traffic associated with construction was unlikely to be significant. Mitigatory measures would therefore not be necessary. In respect of noise impacts from road traffic from the operational phase it was considered that appropriate restrictions were applied such that heavy goods vehicles did not routinely visit the site before 7.00 am. It was also appropriate to limit the number of vehicles visiting the site between 6.00 and 7.00 am to a maximum number per week and it was suggested that heavy goods vehicle (HGV) movements should be restricted to Monday to Friday, and Saturday mornings.

The layout of the proposed development was such that most items of noise-generating fixed plant during operation were located away from the nearest noise-sensitive receptors. Additionally, mitigatory measures were recommended to reduce noise impacts arising from the operation of the plant to satisfactory levels for the bulk delivery of feed and animal noise.

It was suggested that an additional 6 dB attenuation was required to reduce potential noise impacts from bulk feed delivery and animal noise to acceptable levels. This would be provided by appropriate physical screening of the sites by earth bunds.

Lighting

The Local Plan required new developments to minimise the impact of lighting and loss of visual amenity of the night by the use of sensitive illumination. It was recommended that lighting on-site included the use of low pressure sodium lamps and full covers to minimise adverse impacts.

Environmental consequences

Potential odour and ammonia sources would be situated the maximum distance possible from sensitive receptors. The remote and heavily screened location of the proposed development would result in significant odour and ammonia dispersion. Rigorous house-keeping practices and the enclosed, although well ventilated, nature of livestock buildings would also minimise potential odour emissions. Compliance with odour mitigation measures would minimise the likelihood of odour and ammonia emissions, and thus environmental nuisance associated with these sources was considered unlikely.

Noise arising from the construction of the proposed development would be controlled by a planning Section 60/61 Agreement which would enable the control of noise impacts at potentially affected receptors to acceptable levels. Due to the distance from potential noise sources during the operation, the intervening topography, enclosure of the operation within building units, and proposed mitigation measures, it was considered that the specific noise sensitive receptors identified would be exposed to no significant noise impacts resulting from the operation of the proposed development. However, occasional early morning (6.00 – 7.00 am) vehicle movement would need to be controlled on a rota basis.

With the provision of suitable lighting and illumination controls no significant environmental nuisance was envisaged from the construction or operation of the farm.

Waste and water management

An examination of the potential for the development to result in environmental contamination from waste and water management during its construction and operation was undertaken. A review of surface and groundwater features within the vicinity was also undertaken, together with an assessment of on-site drainage requirements and facilities for the management and disposal of effluents and solid organic wastes. There would be no direct discharge of contaminated waters to surface or groundwaters in the construction or operation of the development. It was considered that the construction of the buildings, the staff hostel and associated infrastructure and roadworks had the potential to result in impacts to surface and groundwater in the absence of appropriate mitigation measures. Potential sources of impact from construction activities included release of suspended solids arising from earthmoving works, ground disturbance from construction vehicles and the erosion of stockpiled materials; oil and fuel contamination from re-fuelling, operation and maintenance of construction vehicles; spillages or leakage of any toxic substances used in building construction or formation activity; and construction staff foul sewage. It was anticipated that these impacts could be maintained to within acceptable levels if mitigation measures specified in HRPB pollution control guidelines for engineering works were applied. Foul sewage would be collected and removed by a registered waste carrier for disposal off-site. It was not envisaged that there would be a need to alter the course of any burn (river) during site construction or disrupt any water supply lines in the area. Any burn potentially disturbed during construction works would be culverted in accordance with HRPB requirements.

Aqueous emissions associated with the operation of the development would be suitably managed to avoid unacceptable impacts on surface or groundwater quality by appropriate design of the drainage system. At all times there would be effective segregation of clean and dirty waters, thereby avoiding potential contamination of rainfall run-off from the physical building structures. Sources of wastewater arising during the operation of the pig farm would include: animal slurries; pen flushing water; contaminated surface water; and 'domestic' wastewater. For the purposes of drainage management, emission sources were categorised as Class 1: uncontaminated discharge waters; Class 2: mildly contaminated discharge waters; Class 3: moderately contaminated discharge waters; and Class 4: heavily contaminated discharge waters.

Organic waste management would include the storage and removal of 8,000 tonnes of solid muck (i.e. slurry/straw) per annum and 35,000 tonnes of slurry (including Class 4 wastewaters) per annum. A semi-quantitative risk assessment was undertaken with respect to slurry application to forestry soils, where factors such as heavy metal concentration of slurry, soil pH and bulk permeability are important factors in relation to contamination risk of soils and watercourses, including groundwater.

Inorganic waste materials produced on the farm would include packaging and containers of products purchased and brought onto the farm, and worn out and used materials which are used in the servicing of agricultural machinery (e.g. oils, solvents and detergents). It was recommended that inorganic waste generation during the operation of the farm was minimised by the implementation of waste reduction, re-use and recycling methods and that PIC liaised with the local Waste Disposal Authority (WDA) regarding recycling opportunities for waste produced on-site. It was also recommended that no burning of waste be allowed on-site and that all solid wastes be securely stored prior to collection by the WDA or recycling contractor. No hazardous or special wastes would be generated on site, and any wastes generated from sources not identified above would be treated as controlled waste in accordance with Part II of the Environmental Protection Act (EPA) 1990 and the Duty of Care for Waste contained in section 34 of the Act, and removed by a registered waste carrier.

Environmental consequences

It was envisaged that with the implementation of the outline drainage and storage facilities recommended for water management, together with the implementation of good operational practices there was an insignificant contamination risk to water resources in the vicinity of the application site.

Organic waste recycling to local agricultural and forestry land was considered feasible based on the assessment undertaken by the Agricultural Development Advisory Service ADAS and Land Feed Ltd. Prior to application the suitability of land for slurry and solid muck application would be determined in conjunction with regulatory authorities to ensure that contamination risks are minimised. Operational procedures for the application of wastes would be undertaken according to best agricultural practice, resulting in minimal environmental risk.

Traffic and access

Traffic generated as a result of the construction and operation of the development had the potential to impact upon the local highway network in terms of creating additional volume of movement around road access points. An assessment was therefore undertaken of available information to determine the significance of traffic generation on local roads and access requirements. The Highways Department of HRC was consulted.

There were two roads which would be affected by site related traffic movements. These roads included the B861 Inverness road, where traffic would access the site at the northern corner of the western boundary of the application site adjacent to the proposed site for the staff hostel. This access would be used for operational staff traffic and emergency vehicles and the Daviot Road linking the junction of Balnafoich to the A9 via the bypass of the settlement of Daviot. Construction and operational traffic would access this road via a junction on the south eastern boundary of the application site opposite the Mains of Faillie. Existing traffic volumes on these roads are very low, consisting mainly of local residential and agricultural traffic. Additional traffic volume was associated with the existing quarry operation to the west of Daviot. Quarry traffic was known to access the A9 via the Daviot by-pass thereby avoiding residential areas and a school.

HRC indicated that the Daviot Road junction with the A9 was of suitable standard for access of construction and operational traffic. Daviot Road itself to the point of the existing quarry operation was also of a suitable standard for accommodating vehicles. To the west of the quarry site to the proposed access, the road would require upgrading to accommodate site traffic. Upgrading would encompass the provision of additional passing places and the improvement of existing ones. The upgrading works were not anticipated to result in significant traffic disruption, given the low volumes of vehicles using the local road network, and could be accommodated within the existing road reserve without the need for acquisition of additional land. The volume of staff vehicles accessing Drummossie Farm on the B861 was not anticipated to result in a significant impact on existing road capacity.

Environmental consequences

The anticipated traffic generation associated with the construction and operation of Drummossie Farm was not anticipated to result in significant impacts to the local road network in terms of either volume or access arrangements. The required upgrading works to the Daviot Road would not result in significant traffic disruption.

Ecology

An ecological survey was undertaken following consultation with SNH regarding the location of nearby designated conservation areas (including Sites of Special Scientific Interest (SSSIs)) and Scottish Wildlife Trust (SWT) regarding the ecological value of the application site. SNH indicated that the closest SSSIs to the application site included two sites which were located to the south and east of the proposed development beyond the River Nairn, at about 0.5 and 1.0 km distant respectively. The two sites were known as the Littlemill Fluvio-glacial Landforms and were designated for their glacial deposition features and associated ecology. Available information obtained from SWT indicated that the commercial forestry plantation in which the application site is situated was likely to contain several protected species including badgers, red squirrels and bats.

It was considered that the dense nature of the commercial coniferous plantation currently limits the ecological value of the application site, where ground vegetation was dependent on the maturity of the conifers present. The ground vegetation, including heathers, rushes and mosses, is favoured by more open conditions and it could be expected that the diversity of associated fauna would also be greater in these areas. The vegetation beyond the south-western boundary of the application site, around Loch Caulan, had a higher diversity of vegetation owing to the absence of coniferous trees. No impact on the ecology of the Loch or surrounding habitat, beyond the application boundary, was anticipated. No colonies of bats or badger sets were noted during the habitat survey, although it was recommended that prior to site construction SWT be contacted to confirm that no disturbance of protected species would take place. It was anticipated that the partial clearance of forest within the application boundary for building purposes would result in the creation of a more open habitat and enhanced opportunity for the development of heathland/moorland habitat. The creation of mixed habitat edge between the remaining coniferous forest and the open areas may also enhance the foraging potential for forest fauna and, hence, ecological diversity.

The implementation of HRPB pollution control guidelines for civil engineering works would prevent any contamination of surface water courses and any consequential impacts on aquatic ecology of the area. Following the construction phase, it was recommended that open areas within the application site be maintained for conservation purposes and that the use of pesticides on site other than for cleansing of pig units or infection control be strictly controlled.

Environmental consequences

It was considered that impacts associated with partial clearance of woodland for construction purposes would have minimal impact on the existing ecology of the area, mainly because of the low intrinsic ecological value of coniferous plantation. The creation of open areas within the application site and the enhancement of the interface between remaining forest and open habitat represented significant opportunities to enhance the ecological value of the local area. Furthermore, it was anticipated that the amenity value of the area would be enhanced by the creation of suitable access and walkways for the general public. It was recommended that interpretative information for the general public on any ecological or educational features (e.g. archaeological remains) in the area be provided along access routes. In conclusion, it was considered that the development would be of no adverse significance to the existing habitat within the application site, whereas opportunities exist for enhancing the ecological and amenity value of the site.

Archaeology

The Regional Archaeologist of HRC was consulted with regard to the presence of archaeological remains within, or in the vicinity of, the application site. It was apparent from a review of available information by the Regional Archaeologist that there were archaeological features of interest within the application site boundary. There were ten separate known features present. The majority of these archaeological features were enclosed within the existing conifer plantation as a result of insensitive planting practices when the plantation was established.

The locations of the known archaeological features resulted in the following mitigation steps. Buildings within the three site areas (i.e. pig units sites 1 & 2 and the staff hostel) were located so as to avoid all known archaeological remains and the acknowledgement of the known archaeological remains within the outline Landscape Masterplan with regard to provision of access and opening of the landscape setting. It was considered that by the removal of enclosing forest around these features and the provision of public access (where permissible) to the sites, the amenity and cultural heritage interest of the local area might be significantly enhanced.

During construction it would be necessary to protect the archaeological remains from any site activity or construction plant. It was recommended that measures be adopted on site to ensure that all activities and vehicle movements avoid any archaeological remains. If unrecorded archaeological remains were encountered during excavation activity it was recommended that the Regional Archaeologist be contacted immediately. It was acknowledged that there was the potential for further unrecorded archaeological remains to be present on-site and, upon the recommendation of the Regional Archaeologist and with the support of the developer, a field archaeological survey was proposed.

Environmental consequences

The impact of the construction and operational phases on the known archaeological remains within the application site was anticipated to be insignificant. The removal of surrounding coniferous trees around the archaeological remains, and the provision of public access to the sites (wherever possible) would enhance the amenity and cultural heritage value of the area. It was intended that any other features of archaeological interest identified during the field survey would be suitably recorded and protected.

Conclusion

The ES was submitted to HRC on completion, but, due to strong opposition from opponents to the development the company decide to withdraw their application for planning to minimise any adverse publicity to the company themselves or the pig breeding industry.

Acknowledgements

The authors thank Aspinwall & Company, Locate in Scotland and Egger Ltd. for their kind permission in allowing them to use materials presented in case study one. They also thank Aspinwall & Company for their kind permission in allowing them to use materials presented in example two. The comments made in the text reflect the authors' opinions only.

The final word

Throughout this text many issues are covered relating to a very broad and diversified topic – environmental assessment. The purpose of this book has been not to provide a comprehensive guide to the subject but to show how it can be practically applied at both a project and strategic level. While the book presents examples and case studies from a predominantly British context, it is hoped that it will be of value to readers in other countries.

The development of environmental assessment has been dramatic and interesting. Although environmental assessment is internationally well established as a part of development planning, used by government agencies, companies and other organisations, there is still a need for it to be more fully integrated into the decision-making process. All too frequently, environmental considerations take second place.

The development from project-level environmental assessment to strategic environmental assessment will further widen its application. It is hoped that, in the quest for a sustainable environment, environmental assessment will have an important part to play.

Further reading

A detailed supporting reference list follows this section. However, the reader may also wish to consider the following texts to gain a wider appreciation of the subject of EA.

Ahmad, Y. J. and Sammy, G. K. (1985) Guidelines to environmental impact assessment in developing countries, Hodder & Stoughton, London.

Asian Development Bank (1986) Environmental planning and management, Asian Development Bank.

Burkhardt, D. F. and Ittelson, A. H. (eds) (1978) Environmental Assessment of Socio-economic Systems, Plenum, New York.

Calow, P. (ed.) (1997) Handbook of Environmental Risk Assessment and Management, Blackwell Science, Oxford.

Clark, B. D., Chapman, K., Bisset, R., Wathern, P. and Barrett, M. (1981) A Manual for the Assessment of Major Development Proposals, HMSO, London.

Environment Agency (1996) Environmental Assessment: scoping handbook for projects, Environment Agency, HMSO, London.

Fortlage, C. A. (1990) Environmental Assessment, Gower Publishing Company Ltd.

Glasson, J., Therives, R. and Chadwick, A. (1994) Introduction to Environmental Impact Assessment, UCL Press, London.

Morgan R. K. (1997) Environmental Impact Assessment: a methodological approach, Chapman and Hall, London.

Therival, R. and Patidario, M. R. (1996) The Practice of Strategic Environmental Assessment, Earthscan, London.

Vanclay, F. and Bronstein, D. A. (1995) Environmental and Social Impact Assessment, Wiley, Chichester.

Wood C. (1996) Environmental Impact Assessment: a comparative review, Addison Wesley, Longman, London.

References

Aberdeen Centre for Land Use (1989) *North West Ethylene Pipeline: phase II ecological survey*, ACLU, Aberdeen.

Aberdeen University Research and Industrial Services (AURIS) (1992) *Third River Don Crossing Scoping and Screening Study*, AURIS, Aberdeen.

American Public Health Association (1980) *Standard Methods for the Examination of Water and Wastewater*, WPCF, AWWA.

Andrews, J. and Kinsman, D. (1991) *Gravel Pit Restoration for Wildlife*, Royal Society for the Protection of Birds, Sandy.

Anon. (1963) *The Final Report of the Committee on the Problem of Noise*, Comnd. 2056, Noise, HMSO.

—— (1981) *River Quality: the 1980 Survey and Future Outlook*, National Water Council.

APCOA (1992) *Air Toxic 'Hot Spots' Program Risk Assessment Guidelines*, California Air Pollution Control Officers Association, CA.

Appleby, C. E. (1991) Monitoring at the County Level, in Goldsmith, F. B. (ed.) *Monitoring for Conservation and Ecology*, pp. 155–178, Chapman and Hall, London.

Asian Development Bank (1993) *Environmental Assessment Requirements and Environmental Review Procedures of the Asian Development Bank*, ADB, Manila, Philippines.

Aspinwall & Company (1994a) *Environmental Statement of a Proposed Waste to Energy Plant*.

—— (1994b) *Environmental Statement of a Proposed Sewage Sludge Incinerator – Duncrue Sewage Treatment Works Belfast*.

—— (1995) *Environmental Statement for a Proposed Chipboard Plant, South Gailes, Ayrshire*.

—— (1996a) *Environmental Statement for a Proposed Chipboard Plant, Barony, Ayrshire*.

—— (1996b) *Environmental Statement of Pig Breeding Centre, Drummossie Muir*.

Au, W. K. E. (1994) *Towards Policy and Program EA in Hong Kong*, Paper presented at the Canada–Hong Kong EIA Workshop, Hong Kong, 7–10 March.

Baillie, S. R. (1991) Monitoring Terrestrial Breeding Bird Populations, in Goldsmith, F. B. (ed.), *Monitoring for Conservation and Ecology*, London, Chapman and Hall, pp. 112–132.

Ball, S. and Bell, B. (1991) *Environmental Law*, Blackstone Press Ltd, London.

Batten, L. A., Bibby, C. J., Clement, P., Elliot, G. D. and Porter, R. F. (1990) *Red Book Data Birds in Britain*, Poyser, London.

Beanlands, G. E. and Duinker, P.N. (1983) *An Ecological Framework for Environmental Impact Assessment in Canada*, Institute for Resource and Environmental Studies, Dalhousie University and Federal Environmental Review Office.

Bedfordshire County Council Planning Department (1995) *Bedfordshire Structure Plan 2011 Technical Report 6: Environmental Appraisal*, Bedfordshire County Council Planning Department, Bedford.

Bedfordshire County Council/RSPB (1996) *A Step by Step Guide to Environmental Appraisal*, Bedfordshire County Council, Bedford.

Benington, R. (1993) *Determining Significance in Urban Development Projects in England and Wales*, Unpublished thesis, University of Wales, Aberystwyth.

Bingham, C. S. (1992) *EIA and Project Management: A Process of Communication*, Paper presented at 13th International Seminar on Environmental Impact Assessment, CEMP, Aberdeen.

Bisset, R. (1992) *Environmental Impact Assessment: process, methods and uncertainty*, Paper presented at 13th International Seminar on Environmental Impact Assessment and Management, CEMP, Aberdeen.

Bosanquet, C. H. (1957) The Rise of a Hot Waste Gas Plume, *J. Inst. Fuel*, 30, 197, 322–328.

Bosanquet, C. H., Carey, W. F. and Halton, E. M. (1950) Dust from Chimney Stacks, *Proc. Inst. Mech. Engr.*, 162, 355–367.

Brady, N. C. (1990) Chapter 4 – Physical Properties of Mineral Soils, in *The Nature and Properties of Soils*, 10th Edition, Macmillan, London.

Bratton, J. H. (ed.) (1991) *British Red Data Books: 3. Invertebrates other than Insects*, Joint Nature Conservation Committee, Peterborough.

Briggs, G. A. (1969) *Plume rise, USAEC Critical Review Series*. TID-25075, National Technical Information Service, Springfield, VA.

Burdge, R. J. and Vanclay, F. (1996) Social Impact Assessment – A Contribution to the State of the Art Series, *Impact Assessment*, 14(1).

Cada and Hunsaker (1990) Cumulative Impacts of Hydropower Development: reaching a watershed in impact assessment, *The Environmental Professional* 12(1), 2–8.

Caldwell, L. K., Bartlett, R. V., Parker, D. E. and Keys, D. L. (1982) *A Study of Ways to Improve the Scientific Content and Methodology of Environmental Impact Analysis*, Advanced Studies in Science, Technology and Public Affairs, School of Public and Environmental Affairs, Indiana University, Bloomington.

Canadian International Development Agency (undated) *Guide to Integrating Environmental Considerations*, CIDA, Ottawa.

Canter, L. W. (ed.) (1983) *Water Pollution Impacts in Environmental Impact Assessment*, PADC Environmental Impact Assessment and Planning Unit, Martinus Nijhoff Publishers, The Hague.

—— (1989) Case Study including Management Techniques: The Co-operative Imported Fire

Ant Control Program in the South Eastern USA, Paper presented at the 10th International Seminar on EA, July 1989, University of Aberdeen.

—— (1990) Prediction and Assessment of Noise Impacts, Presented at Second International Course on Environmental Impact Assessment and Management (Module I, Advanced Course on EIA Methods and Techniques), Bologna, Italy, 10–14 December.

—— (1991a) Prediction and Assessment of Groundwater Impacts, Presented at Second International Course on Environmental Conflicts and Impact Assessment (Module III, Advanced Course on EIA Methods and Techniques), Bologna, Italy, April.

—— (1991b) Interdisciplinary Teams in Environmental Impact Assessment, *Environmental Impact Assessment Review* 2 (4) 375–387.

—— (1992) Advanced Environmental Assessment Methods, Paper presented at the 13th International Seminar on Environmental Assessment and Management, Centre for Environmental Management and Planning, Aberdeen.

—— (1996) Environmental Impact Assessment, (2nd edn), McGraw-Hill Inc., New York.

Carruthers, D. (1995) Atmospheric Dispersion Modelling System, Paper presented at Cambridge Environmental Research Consultants Ltd, National Society for Clean Air 1995 Spring Workshop, 29–30 March.

Castro-Morales, L. and Gorzula, S. (1986) The Interrelations of the Caroni River Basin Ecosystems and Hydroelectric Power Projects, *Interciencia*, 11(6), 272–277.

Centre for Environmental Management and Planning (1983) *Post Development Audits to Test the Effectiveness of Environmental Impact Prediction Methods and Techniques*, CEMP, Aberdeen.

—— (1989) *Gold Mining in Connemara and South Mayo – Ireland*, CEMP, Aberdeen.

—— (1994) *Proceedings of a Policy Think-tank on the Effectiveness of Environmental Assessment*, CEMP, Aberdeen.

Clark, B. D., Chapman, K., Bisset, R., Wathern, P. and Barrett, M. (1981) *A Manual for the Assessment of Major Developments Proposals*, HMSO, London.

Cleland, D. I. and Kerzner, H. (1986) *Engineering Team Management*, Van Nostrand Reinhold Company, New York.

Coggon, D. (1995) Epidemiology and the Assessment of Small Risks, Trans. *IChemE*, Part B, 73(B4), S36–S38.

Coles, T. F. and Tarling, J. P. (1991) *Environmental Assessment: Experience to Date*. Institute of Environmental Assessment.

Commission of the European Communities (1976) *Environment Programme 1977–1981*.

—— (1990) *Proposal for Directive on the Environmental Assessment of Policies, Plans and Programmes*, CEC XI/194/90 – Rev. 1-EN. Brussels, 16 August.

—— (1992) *Towards Sustainability: a community programme of policy and action in relation to the environment and sustainable development*, (The EC Fifth Environment Action Programme). CEC Official Publication of the EC (CM (92) 23/11 Final), CEC, Brussels, Belgium.

—— (1993) *Report of the Commission on the implementation of Directive 85/337/EEC on assessment of the effects of certain public and private projects on the environment*, COM (93) 28 final. Vol. 12.

—— (forthcoming) *Proposal for a Council Directive on the assessment of the effects of certain plans and programmes on the environment*.

Coppin, N. J. and Bradshaw, A. D. (1982) *Quarry Reclamation, The Establishment of Vegetation in Quarries and Open Pit Non-metal Mines*, Mining Journal Books, London.

Cordah (1997) *QA Procedures Forum*, Personal Communication.

Council for the Protection of Rural England (CPRE) (1990) *Environmental Statements: Getting Them Right*, Council for the Protection of Rural England, London.

—— (1991) *Review of the Town and Country Planning Inquiries Procedure Rules*, CPRE, London.

Countryside Commission (1987) *Landscape Assessment: A Countryside Commission Approach*, CCD 18, Countryside Commission, Cheltenham.

—— (1988) *A Review of Recent Practice and Research in Landscape Assessment*, Landscape Research Group, Commissioned by the Countryside Commission, Cheltenham.

—— (1991) *Environmental Assessment: the treatment of landscape and countryside recreation issues*, CCP 326. Countryside Commission, Cheltenham.

Culhane, P. J. (1987) The Precision and Accuracy of U.S. Environmental Statements. *Environmental Monitoring and Assessment* 8, 217–238.

Davidson, W. F. (1949) The Dispersion and Spreading of Gases and Dust from Chimneys, *Trans Conf. on Ind. Wastes*, 14th Ann. Meeting Industrial Hygiene Foundation of America, 38–55.

Department of Environment Affairs (1992a) *The Integrated Environmental Management Procedure*, Department of Environment Affairs, Pretoria, Republic of South Africa.

—— (1992b) *Guidelines for Scoping*, Department of Environment Affairs, Pretoria, Republic of South Africa.

—— (1992c) *Guidelines for Report Requirements*, Department of Environment Affairs, Pretoria, Republic of South Africa.

—— (1992d) *Guidelines for Review*, Department of Environment Affairs, Pretoria, Republic of South Africa.

—— (1992e) *Checklist of Environmental Characteristics*, Department of Environment Affairs, Pretoria, Republic of South Africa.

Department of the Environment UK (1986) *Waste Management Paper No. 26, Landfilling Wastes*, HMSO, London.

—— (1988) *Town and Country Planning (Assessment of Environmental Effects) Regulations (SI No. 1199)*, London, HMSO.

Department of the Environment/Welsh Office UK (1989) *Environmental Assessment: a guide to the procedures*, HMSO, London.

Department of the Environment UK (1991) *Policy Appraisal and the Environment*, HMSO, London.

—— (1992a) Chief Inspector's Guidance to Inspectors, Environmental Protection Act 1990, *Process Guidance Note IPR 5/3*, Waste Disposal & Recycling Municipal Waste Incineration, HMSO, London.

—— (1992b) *Planning Policy Guidance Note 12: Development Plans and Regional Planning Guidance*, HMSO, London.

—— (1993) *Environmental Appraisal of Development Plans: A Good Practice Guide*, HMSO, London.

—— (1994a) *Sustainable Development: The UK Strategy*, HMSO, London.

—— (1994b) *Environmental Appraisal in Government Departments*, HMSO, London.

—— (1995) *A Guide to Risk Assessment and Risk Management for Environmental Protection*, HMSO, London.

—— (1997) *UK Air Quality Standard Regulations*, London, HMSO.

Department of Natural Resources (1994) *Tourism and Environment in the Victoria Falls Area. An assessment of the Environmental Impact of Tourism Developments*, Phase 1 Study Report. Department of Natural Resources, Harare, Zimbabwe.

Department of Transport UK (1983) *Manual of Environmental Appraisal*, Department of Transport, London, HMSO.

—— (1992) *Assessing the Environmental Impact of Road Schemes. Report from the Standing Advisory Committee on Trunk Road Assessment (SACTRA)*, HMSO, London.

—— (1994) *Design Manual for Roads and Bridges, Volume II Environmental*. Department of Transport, London.

de Vries, Y. (1996) The Netherlands experience. *Environmental Assessment of Policies-Briefing Papers on Experiences in Selected Countries*, J. J. de Boer and B. Sadler (ed.) Paper No. 54, Ministry of Housing, Spatial Planning and the Environment, The Netherlands, pp. 67–74.

Deutsche Gesellschaft für Technische Zusammenarbeit (GTZ) GmbH (1988) ZOPP (an introduction to the method). GTZ, Frankfurt am Main, Germany.

DHV Environment and Infrastructure (1994) *Strategic Environmental Assessment, Existing methodology*. Commission of the European Union, Directorate General for Environment, Nuclear Safety and Civil Protection, Brussels, Belgium.

Directorate of Environmental Affairs (1995) *Namibia's Environmental Assessment Policy*, Directorate of Environmental Affairs, Ministry of Environment and Tourism, Windhoek, Namibia.

Dixon, J. (1994) *Strategic Environmental Assessment: the New Zealand Experience*, Paper presented at IAIA'94. Quebec City, Canada. International Association for Impact Assessment.

Doll, R. (1995) Assessment of Risk from Low Doses: Contribution of Epidemiology, *Trans. IChemE*, Part B, 73(B4), S8–S11.

Dony, J. G. and Denholm, I. (1985) Some Quantitative Methods of Assessing the Conservation Value of Ecologically Similar Sites, *Journal of Applied Ecology*, 22, 229–238.

Dorney, R. S. and Dorney, L.C. (1989) *The Professional Practice of Environmental Management*, Springer

EDELCA (1987) *El hombre, destinario de todo desarrollo*, EDELCA, Revista de CVG/Electrificación del Caroni C.A., 5: 6–11.

Ellis, J. B., (1976) Sediments and Water Quality of Urban Stormwater, *Wat. Services* 80, 730–734.

—— (1984) Personal Communication, Urban Pollution Research Centre, Middlesex Polytechnic, London, UK.

Emery, M. (1986) *Promoting Nature in Cities and Towns: a practical guide*, Croom Helm, London.

Environment Canada (1974) *An Environmental Assessment of Nanaimo Post Alternatives*, Environment Canada, Ottawa.

Environmental Protection Agency (1995) *Environmental Impact Assessment Procedures*, Environmental Protection Agency, Accra, Ghana.

Environmental Resource Management (1993) *Environmental Statement for the Shell Green Sewage Sludge Incinerator for North West Water Ltd.*

Environmental Resources Ltd (1987) *South Warwickshire Prospect: Environmental Impact Assessment*, ERL, London.

Espoo Convention (1991) Convention on Environmental Impact Assessment, Finland, February.

ESSA Technologies Ltd (1994) *Benefits of Environmental Assessment*, ESSA Technologies Ltd, Ottawa, Ontario, Canada.

European Bank for Reconstruction and Development (1994) *Environmental Impact Assessment Legislation: Czech Republic, Estonia, Hungary, Latvia, Lithuania, Poland, Slovak Republic, Slovenia*, European Bank for Reconstruction and Development, London.

European Communities (1997) Council Directive 97/11/EC of 3 March amending Directive 85/337/EEC on the assessment of the effects of certain public and private projects on the environment, *Official Journal of the European Communities* No L 73/5, 14.3.97.

European Council Directive (1985) On the assessment of the effects of certain public and private projects on the environment, 85/337/EC, *Official Journal of the European Communities*, 27 June.

Eyre, M. D., Rushton, S. P., Luff, M. L., Ball, S. G., Foster, G. N. and Topping, C. J. (1986) *The Use of Invertebrate Community Data in Environmental Assessment*, Agricultural Environment Research Group, University of Newcastle upon Tyne.

Eyre, M. D. and Rushton, S. P. (1989) Quantification of Conservation Criteria using Invertebrates, *Journal of Applied Ecology* 26, 159–171.

Federal Environmental Assessment Review Office (FEARO) (1986) *Initial Assessment Guide*, FEARO, Ottawa.

—— (1992) *Environmental Assessment in Policy and Program Planning, A Source Book*, FEARO, Ottawa, Canada.

Fergusson, C. and Marsh, J. (1993) Assessing Human Health Risks from Ingestion of Contaminated Soil, *Land Contamination and Reclamation* 1(4), 177–185.

Fletcher, I. J., Pratt C. J. and Elliot G. E. P. (1978) An Assessment of the Importance of Roadside Gulley Pots in Determining the Quality of Stormwater Runoff, *Proc. Int. Conf. on Urban Storm Drainage*, Helliwell, P.(ed.), Pentech Press, London, 586–602.

Goldsmith, E. and Hildyard, N. (eds) (1984) *The Social and Environmental Effects of Large Dams, Volume 1: Overview*, Wadebridge Ecological Centre, Cornwall.

—— (1986) *The Social and Environmental Effects of Large Dams, Volume 2: Case Studies*, Wadebridge Ecological Centre, Cornwall.

Goldsmith F. B. (ed.) (1991) *Monitoring for Conservation and Ecology*, Chapman and Hall, London.

Gorzula, S. (1992) Social Impact Assessment, Paper presented at the 7th International Training Course in Environmental Assessment and Management. Centre for Environmental Management and Planning, Aberdeen.

Goss-Custard, J. D. *et al.* (1991) Towards Predicting Wading Bird Densities from Predicted Prey Densities in a Post-Barrage Severn Estuary, *Journal of Applied Ecology*, 28, 1004–1026.

Gotts, R. E. (1993) *Toxic Risks: Science, Regulation and Perception*, Lewis Publishers, Chelsea, MI.

Gow, L. J. A. (1994) The New Zealand Experience in Policy Environmental Assessment, Paper to the International Workshop on Policy Environmental Assessment, The Hague, Netherlands, 12–14 December.

Grampian Regional Council (1995) *Grampian Structure Plan: Environmental Appraisal*, Grampian Regional Council, Aberdeen.

Gregory, K. J. and Walling, D. E. (1970) The Measurement of the Effects of Building and Construction on Drainage Basin Dynamics, *Journal of Hydrology*, 11, 129–144.

Harding, P. T. (1991) National Species Distribution Surveys, in Goldsmith F. B. (ed.) *Monitoring for Conservation and Ecology*, pp. 133–154, Chapman and Hall, London.

Harrop, D. O. (1984) *Urban Pollution Research Centre, Report 7*, Middlesex Polytechnic, London, UK, April, 110pp.

—— (1994) Environmental Assessment and Incineration, in Harrison, R. M. and Hester, R. (eds), *Waste Incineration and the Environment*, Royal Society of Chemistry, London.

—— (1999) Air Quality Assessment, in Petts J. *et al.* (eds), *Handbook on Environmental Impact Assessment*, Blackwell, in press.

Harrop, D. O. and Carpenter, R. P. (1992) *Methods for Assessing Air Quality Impact*, Proceedings of the 59th Conference of the National Society for Clean Air and Environmental Protection, Bournemouth, England.

Harrop, D. O. and Pollard, S. (1998) Quantitative Risk Assessment for Incineration: Is it Appropriate for the UK?, *J. CIWEM*, 12, 48–53.

Hawley, J. K. (1985) Assessment of the Health Risks from Exposure to Contaminated Soil, *Risk Analysis* 5, 289–302.

Health and Safety Executive (1989) *Risk Criteria for Landuse Planning in the Vicinity of the Major Industrial Hazards*, HMSO, London.

Holland, J. Z. (1953) *A Meteorological Survey of the Oak Ridge Area*, Atomic Energy Comm., Report ORO-99, Washington, D.C.

Hollis, G. E. and Luckett J. R. (1976) The Response of Natural River Channels to Urbanisation: Two Case Studies from South East England, *Journal of Hydrology*, 30, 351–363.

Hrudey, S. E. and Krewski, D. (1995) Is There a Safe Level of Exposure to a Carcinogen?, *Environmental Science and Technology*, 29(8), 370A–375A.

Institute of Chemical Engineering (1994) Prediction and Assessment Techniques, Chapter 8, in *Environmental Training for the Process Industries*, Safety, Health and Environment Department, Institute of Chemical Engineers, London.

Institute of Environmental Assessment (1995) *Guidelines for Baseline Ecological Assessment*, E & FN Spon, London.

Institute of Environmental Assessment and the Landscape Institute (1995) *Guidelines for Landscape and Visual Impact Assessment*, E & FN Spon, London.

International Chamber of Commerce (1991) *Business Charter for Sustainable Development*, London.

International Organisation for Standardisation (1996) *Environmental Management Systems-Specification with Guidance for Use*, International Organisation for Standardisation, Geneva, Switzerland.

International Standards Organisation (1985) ISO 7828, *Water Quality – Methods of Biological Sampling: Guidance on Hand Net Sampling of Benthic Macroinvertebrates*, International Standards Organisation.

IUCN Regional Office for Southern Africa (1994) *Victoria Falls: Strategic Environmental Assessment*, Scoping Workshop Report. IUCN, Gland, Switzerland.

Jones, C. E. and Lee, N. (1993) *Post-auditing in Environmental Assessment: the Greater Manchester Metrolink Scheme*, Occasional Paper 37, EIA Centre, University of Manchester.

Kelly, K. (1991) The Myth of 10^{-6} as a Definition of 'Acceptable Risk', proceedings 84th Annual Meeting of the Air and Waste Management Association, Vancouver, BC, Canada, pp. 2–10.

Kent, M. and Coker, P. (1992) *Vegetation Description and Analysis: a practical approach*, Bellhaven Press, London.

Kent County Council (1993) *Strategic Environmental Appraisal of Policies*, Working Paper No. 1/93 to Kent Structure Plan, Kent County Council.

Kramer L. (1990) *EEC Treaty and Environmental Protection*, Sweet & Maxwell, London.

Lambrick, G. (1993) Environmental Assessment and Cultural Heritage: Principles and Practice, in Ralston, I. and Thomas, R. (eds) Environmental Assessment and Archaeology, Occasional Paper No. 5, Institute of Field Archaeologists, June.

Law, R. (1994) Environmental Assessment of Policy Initiatives, The Hong Kong Experience, Paper presented to the International Workshop on Policy Environmental Assessment, The Hague, Netherlands, 12–14 December.

Le Blanc, P. and Fisher, K. (1994) Application of Environmental Assessment to Policies and Programmes: the Federal Experience in Canada, Paper presented to the International Workshop on Policy and Environmental Assessment, The Hague, Netherlands, 12–14 December 1994.

Lee Wilson and Associates, Inc. (1992) *Verification of NEPA Predictions on Environmental Assessments of Oklahoma Surface Coal Mines*, Report prepared for US Environmental Protection Agency, Region 6, Dallas, Texas.

—— (1993) *Imparting Region 6 EISs and NEPA Decision-making, Task 2 Report: Protocols for Post-EIS Monitoring and Cumulative Impact Assessment*, Report prepared for US Environmental Protection Agency, Region 6, Dallas, Texas.

—— (1995) *Impacts-backwards Method for Doing EIA Audits*, Draft report, Santa Fe, NM.

Lee, N. (1991) Quality Control in Environmental Assessment, in *Proceedings* of Advances in Environmental Assessment Conference, London, 28–29 October.

Lee, N., Perry, R. and Ford, R. (1983) *A Rail Strategy for Greater Manchester: Environmental Evaluation of the Major Options*, Greater Manchester Passenger Transport Executive (GMPTE), Manchester.

Lee, N. and Colley, F. (1990) *Reviewing the Quality of Environmental Statements*, Occasional Paper 24, Department of Planning and Landscape, University of Manchester.

Lee, N. and Colley, F. (1992) *Reviewing the Quality of Environmental Statements* (second edition) Occasional Paper 24, Department of Planning and Landscape, University of Manchester.

Lee, N. and Walsh, F. (1992) Strategic Environmental Assessment: An Overview, *Project Appraisal* 7 (3), 126–136.

Leopold, L. B., Clarke, F. E., Kanshaw, B. B. and Balsley, J. R. (1971) *A Procedure for Evaluating Environmental Impact*, US Geological Survey Circular No. 654, US Geological Survey, Washington DC.

Malloch, A. J. C. (1990) *MATCH: a computer program to aid the assignment of vegetation data to the communities and sub-communities of the National Vegetation Classification*, University of Lancaster.

Mance, G. and Harman M. I. I. (1978) The Quality of Urban Storm-water Run-off, *Proceedings of an International Conference on Urban Storm Drainage*, Helliwell P. (ed.), Pentech Press, London, 603–618.

Martin, J. (1985) Landscape Evaluation and Visual Impact Assessment: An Overview, Paper presented at the 6th International Seminar on EIA, CEMP, Aberdeen.

Martin, K. (1984) Visual Impact Analysis of Mining and Waste Tipping Sites – A Review, Paper presented at Strategies for Environmentally Sound Development in the Mining and Energy Industries, CEMP, Aberdeen.

Maynard, K. M., Cameron, R., Fielder, A., McDonald, A. and Wadge, A. (1995) Setting Air Quality Standards for Carcinogens: An alternative to mathematical quantitative risk assessment, *Human and Experimental Toxicology*, 14, 175–186.

McHarg, I. (1968) *A Comprehensive Highway Route Selection Method*, Highway Research Record No. 246, Highway Research Board, Washington DC.

—— (1969) *Design with Nature*, Natural History Press, New York.

McQuaid-Cook, J. (1994) Consultation and Participation in Environmental Assessment, Paper presented at a training course on environmental assessment, October to November, CEMP, Aberdeen.

Meynell, P.-J. (1997) Strategic Environmental Assessment: Planning for Cumulative Impacts of Tourism Development, Paper presented at the Environment Matters Conference, An International Conference for the Tourism, Hospitality and Leisure Sectors, Glasgow, 29 April–2 May 1997.

Meynell, P.-J., Nalumino, N. and Sola, L. (1996) Working Towards Cross-border Management of a World Heritage Site: the Strategic Environmental Assessment of Developments around Victoria Falls, Zambia/Zimbabwe, Paper presented at the World Conservation Congress, Montreal, Canada, October.

Ministry of Housing, Spatial Planning and the Environment (1989) *To Choose or to Lose–National Environmental Policy Plan*, Ministry of Housing, Spatial Planning and the Environment, The Hague, Netherlands.

—— (1992) *National Environmental Policy Plan*, Ministry of Housing, Spatial Planning and the Environment, The Hague, Netherlands.

—— (1994) *National Environmental Policy Plan*, Ministry of Housing, Spatial Planning and the Environment, The Hague, Netherlands.

Morrey, J., Aldous, P. J., Quint, M. and Jefferies, S. R. (1996) The Use of Probabilistic Risk Assessment Techniques for Groundwater Resource Management and Groundwater Protection, Paper presented to IBC Groundwater Conference.

Moses, H., Strom, G. H. and Carson, J. E. (1964) Effects of Meteorological and Engineering Factors on Stack Plume Rise, *Nuclear Safety*, 6, 1, 1–19.

Moss, B. (1994) *Sustainable Land Use in the Scottish Uplands*, ITE Draft report, ITE, Banchory.

Mudge G. (1994) Environmental Assessment – Present Problems and Future Prospects as Seen in International Academy of the Environment, *Capacity Building in Environment and Development*, Geneva, Switzerland.

Murley, L. (1995) *Clean Air Around the World*, 2nd ed., IUAPPA, Brighton.

Nair, C. *et al.* (1994) Strategic EIA: The Environmental Assessment of Plans, Policies and Programmes in Hong Kong, Paper presented at the Canada–Hong Kong EIA Workshop, Hong Kong, 7–10 March.

National Physical Planning Agency (1993) *Environmental Impact Assessment for the Lake Myvatn area in Skutustadahreppur*, NPPA, Reykjavik.

National Rivers Authority (1992) *River Corridor Surveys: Methods and Procedures, Conservation Technical Handbook No. 1*, NRA, Newcastle upon Tyne.

National Society for Clean Air and Environmental Protection (1998), *Pollution Handbook*, Brighton.

Nature Conservancy Council (1990) *Handbook for Phase 1 Habitat Survey, A Technique for Environmental Audit*, NCC, Peterborough.

—— (1991) *Site Management Plans for Nature Conservation*, Nature Conservancy Council, Peterborough.

Neufeld, G. (1992) *A Preliminary Survey of the Impact of Environmental Assessments on Competitiveness*, Economic Council of Canada.

Nixon, J. A. (1994) *The Practical Application of Environmental Impact Assessment in Iceland*, papers presented for a training course, 18 – 22 April, Reykjavik.

Norris, K. (1996) The European Commission Experience, Environmental Assessment of Policies-briefing Papers on Experiences in Selected Countries, de Boer, J. J. and Sadler, B. (eds), Paper No. 54, Ministry of Housing, Spatial Planning and the Environment, Netherlands, pp. 51–56.

North West Water (1992) *Proposed Wastewater Treatment Works at Jameson Road, Fleetwood*, Public Inquiry, November, Proof of Evidence.

Organisation for Economic Co-operation and Development (OECD Development Assistance Committee) (1992) *Guidelines on Aid and Environment No. 1: Good Practices for Environmental Impact Assessment of Development Projects*, OECD, Paris.

Oxford Archaeological Unit (1991) *British Railways Board Rail Link Project Eastern Section Environmental Assessment: Specialist Study of Historical and Cultural Impacts*, British Rail.

Pasquill, F. and Smith, F. B. (1982) *Atmospheric Diffusion*, 3rd edn, Ellis Horwood, Chichester.

Perring, F. H. and Farrell, L. (1983) *British Red Data Books: 1. Vascular Plants*, Royal Society for Nature Conservation, Lincoln.

Petts, J. (1994) Incineration as a Waste Management Option, in Hester, R. E. and Harrison, R. M. (eds), *Waste Incineration and the Environment, Issues in Environmental Science and Technology*, Royal Society of Chemistry, 1–25.

—— and Eduljee, G. (1994) *Environmental Impact Assessment for Waste Treatment Disposal Facilities*, John Wiley and Sons, Chichester.

Phillipson, D. W. (ed.) (1990) *Mosi-oa-Tunya: A Handbook to the Victoria Falls Region*, Longman Zimbabwe, Harare.

Pinfield, G. (1992) Strategic Environmental Assessment and Land Use Planning, *Project Appraisal* 7 (3), 157–163.

Pollard, E. (1991) Monitoring Butterfly Numbers, in Goldsmith F. B. (ed.), *Monitoring for Conservation and Ecology*, London, Chapman and Hall, pp. 87 111.

Pollard, S. J. T., Harrop, D. O., Crowcroft, P., Mallett, S. H., Jeffries, S. R. and Young, P. J. (1995) Risk Assessment for Environmental Management: Approaches and Applications, *J. CIWEM*, 9, 621–628.

Price, F. T. (1994) Developing Standards for Environmental Risk Assessment – In Pursuit of Consensus, *ASTM Standardisation News*, May, 27–29.

Purdue, M. (1991) *Planning Appeals: A Critique*, Open University Press, Milton Keynes.

Purnell, S. (1992) *Putting Assessment in the Public Eye*, Environmental Assessment and Audit: A User's Guide.

Ralston I. and Thomas R. (1993) Environmental Assessment and Archaeology: an Introduction, in *Environmental Assessment and Archaeology*, Ralston, I. and Thomas, R. (eds), Occasional Paper No. 5, Institute of Field Archaeologists, June.

Ratcliffe, D. A. (1971) Criteria for the Selection of Nature Reserves, *Advancement of Science*, 27, 294–296.

—— (ed.) (1977) *A Nature Conservation Review*, Volumes 1 and 2. Cambridge University Press, Cambridge.

Raymond, K. (1994) Strategic EA: Putting the Theory into Practice, Paper presented at the 3rd European Workshop of EIA Centres Delphi, Greece.

Rodricks, J. V. (1992a) *Calculated Risks: The toxicity and human health risks of chemicals in our environment*, Cambridge University Press, Cambridge.

—— (1992b) Risk Assessment at Hazardous Waste Disposal Sites, *Haz. Waste* 1 (3), 333–362.

Rodwell, J. S. (ed.) (1991a) *British Plant Communities Volume 1: Woodlands and scrub*, Cambridge University Press, Cambridge.

—— (ed.) (1991b) *British Plant Communities Volume 2: Mires and heaths*, Cambridge University Press, Cambridge.

—— (ed.) (1992) *British Plant Communities Volume 3: Grassland and montane vegetation*, Cambridge University Press, Cambridge.

—— (ed.) (1993) *British Plant Communities Volume 4: Aquatic communities, swamps and tall-herb fens*, Cambridge University Press, Cambridge.

—— (ed.) (1995) *British Plant Communities Volume 5: Maritime and weed communities*, Cambridge University Press, Cambridge.

Royal Commission on Environmental Pollution (1989) *Incineration of Waste*, Seventeenth Report, HMSO, London.

Sadler, B. (1996) *Environmental Assessment in a Changing World: Evaluating Practice to Improve Performance*, International Study on the Effectiveness of Environmental Assessment, Ministry of Supply and Services, Canada.

Sadler, B. and Verheem, R. (1996) *Strategic Environmental Assessment-Status, Challenges and Future Directions*, Paper No. 53, Ministry of Housing, Spatial Planning and the Environment, Netherlands.

Sampson, D. (1992) Landscape and Visual Impact Assessment of Land Reclamation Works, Proceedings of the British Land Reclamation Group Winter Meeting. University of Sheffield, December.

Schaenam, P. S. (1976) *Using an Impact Measurement System to Evaluate Land Development*, The Urban Institute, Washington DC.

Scholten, J. J. (1997) Reviewing EISs – the Netherlands Experience, *Environmental Assessment* 5 (1), 24–25.

Sheate, W. R. (1992) Strategic Environmental Assessment in the Transport Sector, *Project Appraisal*, 7 (3), 170–174.

Shell Chemicals UK Ltd (1989) *Environmental Statement, The North Western Ethylene Pipeline*, Shell, Chester.

Shell International (1994) *Environmental Assessment*, Report EP 94–1980. Shell International, The Hague, Netherlands.

Shirt, D. B. (1987) *British Red Data Books: 2, Insects*, Nature Conservancy Council, Peterborough.

Shopley, J. B. and Fuggle, R. F. (1984) A Comprehensive Review of Current EA Methods and Techniques, *Journal of Environmental Management*, 18, 25–48.

Sielcken, R. J., Scholten, J. J. and van Eck, M. (1996) Review Criteria Employed by the Commission for EIA in the Netherlands, *Proceedings* of the International Association for Impact Assessment Conference, Lisbon, Portugal.

Sigal, L. and Webb, W. (1994) A Case Study of the Application of Strategic Environmental Assessment (SEA) to an Environmental Restoration and Waste Management Program, Paper presented at IAIA '94, Quebec City, Canada. International Association for Impact Assessment.

Sippe, R. (1994) Policy and Environmental Assessment in Western Australia: Objectives, Options, Operations and Outcomes, Paper presented at the International Workshop on Policy and Environmental Assessment, The Hague, Netherlands, 12–14 December.

Smarden, R. C., Palmer, J. F. and Felleman, J. P. (1986) *Foundations for Visual Project Analysis*, John Wiley and Sons.

Southern Water Services Ltd (1991) *Testwood Lakes Environmental Statement*, Southern Water Services Ltd, Hampshire Division.

Spellerberg, I.F. (1992) *Evaluation and Assessment for Conservation: ecological guidelines for determining priorities for nature conservation*, Chapman and Hall, London.

—— (1994) The Biological Content of Environmental Assessments, *Biologist*, 41(3), 126–128.

Steinitz, C., Parker, P. and Jordan, L. (1976) Hand Drawn Overlaps: Their History and Prospective Uses, *Landscape Architecture*, 66, 444–455.

Stephenson, D., Brooke, C. E. and Nixon, J. A. (1995) *Public Participation in EA: a review of experience in Europe and the UK*, CEMP, Aberdeen.

Szepesi, D. J., (1989) *Compendium of Regulatory Air Quality Simulation Models*, Akadémiai Kiadó, Budapest, Hungary.

Tarling, J. P. (1991) A Comparison of Environmental Assessment Procedures and Experience in the UK and the Netherlands, MSc in Environmental Management Dissertation, University of Stirling, Scotland.

Thames Region NRA (National Rivers Authority) (1989) *Environmental Assessment Guidelines: a procedure for ensuring environmental factors are taken into account in the design and implementation of land drainage improvement works*, Thames Region National Rivers Authority, Reading.

Thor, E. C., Elsner, G. H., Travis, M. R. and O'Loughlin, K. M. (1978) Forest Environmental Impact Analysis – a new approach, *Journal of Forestry*, 76 (2), 723–725.

Treweek, J. (1995) Ecological Impact Assessment, in Vanclay, F. and Bronstein, D. A. (eds) *Environmental and Social Impact Assessment*, pp. 171–193, Wiley, Chichester.

213

Tucker, C. G. J. (1975) MSc. Thesis, Department of Civil and Structural Engineering, Trent Polytechnic, Nottingham, UK.

Turnbull, R. (1997) Personal Communication.

Turner D. B. (1970) *Workbook of Atmospheric Dispersion Estimates, Air Resources Field Research Office*, Environmental Science Services Administration, Environmental Protection Agency, Offices of Air Programs, Research Triangle Park, North Carolina.

—— (1979) Atmospheric Dispersion Modelling – A Critical Review, *Journal of Air Pollution Control Association*, 29 (5), 502–519.

—— (1994) *Workbook of Atmospheric Dispersion Estimates: An Introduction to Dispersion Modelling* (2nd edn), Lewis Publishers, Ann Arbor.

United Nations Environment Programme (1992) *Agenda 21*, UNEP, Rio de Janeiro.

United States Environmental Protection Agency (1986) *Superfund Public Health Evaluation Manual*, EPA/540/1–86/60, October.

—— (1987) *Industrial Source Complex (ISC) Dispersion Model User's Guide* (2nd edn) Volumes 1 and 2, EPA-450/4-88-002A and EPA-450/4-88-002B, December. EPA, Washington, DC.

—— (1990) *Exposure Factors Handbook, Office of Health and Environmental Assessment*, EPA/600/8–89/043, US EPA, Washington, DC.

—— (1994) *Estimating Exposure to Dioxin-Like Compounds*, Review Draft, EPA/600/6–88/005 CA, Office of Research and Development, Washington DC.

Usher, M. B. (ed.) (1986) *Wildlife Conservation Evaluation*, Chapman and Hall, London.

Verheem, R. (1992). Environmental Assessment at the Strategic Level in the Netherlands, *Project Appraisal* 7 (3), 150–156.

—— (1994) SEA of Dutch Ten Year Programme on Waste Management, Paper presented at IAIA '94, Quebec City, Canada, International Association for Impact Assessment.

—— (1996) SEA in the Netherlands: Overview and Some Case Studies, Paper presented to the 'Fachtagung Umweltprüfung von Plänen und Programmen', Vienna, Austria, 16 April.

Walsh, F. (1996) Strategic Environmental Assessment of Policies, Paper presented at the 13th International Training Course on Environmental Assessment and Management, July–August, CEMP, University of Aberdeen.

Walsh, F., Lee, N. and Wood, C. (1991) *The Environmental Assessment of Opencast Coal Mines*, Department of Planning and Landscape, University of Manchester.

Wathern, P. (1984). Ecological Modelling in Impact Analysis, in Roberts, R. D. and Roberts, T. M. (eds) *Planning and Ecology*, London, Chapman and Hall, pp. 80–98.

—— (ed.) (1989) *Environmental Impact Assessment: theory and practice*, London, Unwin Hyman.

Webb, J. W. and Sigal, L. L. (1992) Strategic Environmental Assessment in the United States, *Project Appraisal* 7 (3), 137–142.

Westman, W. E. (1985) *Ecology, Impact Assessment and Environmental Planning*, John Wiley & Sons, New York.

Wilkinson, R. (1956) *Journal of the Institution of Public Health Engineers*, Vol. 55, 70–84.

Wilson, P. *et al.* (1996) *Emerging Trends in National Environmental Legislation in Developing Countries*, United Nations Environment Programme, Nairobi, Kenya.

Wood, C. and Jones, C. (1991) *Monitoring Environmental Assessment and Planning*. Department of the Environment (Planning Research Programme), HMSO, London.

World Bank (1991) *Environmental Assessment Sourcebook – Policies, Procedures, and Cross Sectoral Issues*, Technical Paper No. 139. World Bank Environment Department, (Sec. M93–212), Washington, DC.

—— (1993) *Annual Review of Environmental Assessment*, World Bank Environment Department (Sec. M93–212). Washington, DC.

—— (1995a) Presentation on EIA in Africa, Annual Conference of the International Institution for Impact Assessment, Durban, Republic of South Africa, June.

—— (1995b) The Impact of Environmental Assessment, Draft prepared by Land, Water and Natural Habitats Division, World Bank Environment Department, Washington, DC.

World Business Council on Sustainable Development (1995) *Environmental Assessment: a Business Perspective*, World Business Council on Sustainable Development, Geneva, Switzerland.

World Health Organisation (1987a) *Air Quality Guidelines for Europe*, WHO Regional Publication, European Series No. 23, Copenhagen.

—— (1987b) Environmental Health Series 15, *Health and Safety Component of Environmental Impact Assessment*, Copenhagen.

—— (1995a) *Updating and Revision of the Air Quality Guidelines for Europe*, October, Dusseldorf, Germany.

—— (1995b) *Update and Revision of the Air Quality Guidelines for Europe*, October, Bilthoven, Netherlands.

Yeater, M. and Kurokulasuriya, L. (1996) *Environmental Impact Assessment Legislation in Developing Countries*, United Nations Environment Programme, Nairobi, Kenya.

Index

Aberdeen 28–30
additive effects 151
Agenda 21 2, 21, 99
air quality assessment 34–40, 80–82, 182–183, 192–193; air dispersion modelling 37–40; ambient air quality standards (AAQS) 35, 37–38; definition 34; methodology 35
Alder Road Weir 30
alternatives 24
archaeological assessment 64–66, 198–199; methodology 65
Asian Development Bank (ADB) 6
Aswan Dam 5
audit 128, 138–146
Australia 5, 150

baseline studies 18, 176
Bedfordshire 102
Belfast 123–125
best practicable environmental option (BPEO) 123
biological monitoring working group 53
biological records centre (BRC) 53
British Coal 105

Canada 2, 5, 6, 10, 21, 57, 105–107, 119–120, 122, 128, 134–135, 150, 151–152, 154

case studies 28–31, 38–40, 44–45, 49–52, 57–58, 63–64, 66, 69–71, 80–88, 101–107, 123–125, 141–146, 161–169, 171–199
Centre for Environmental Management Planning (CEMP) 3, 102; 141–142
Channel Tunnel 65–66
checklists 16, 19–20; basic checklist 19; descriptive checklist 19–20; scaling checklist 19
chipboard manufacture 44–45, 49–52, 172–184
Connemara 101
consultation 89–107, 176–177, 189
cost benefit analysis (CBA) 5
Council for Environmental Quality (CEQ) 160
Council for the Protection of Rural England (CPRE) 99
Countryside Commission 45–48
Cow Green 141–142
cultural heritage 64–66; methodology 65
cumulative effects assessment 151
cumulative impacts 17, 146, 150, 151–152, 160, 165, 167

dams 69–71, 151–152, 161
descriptive matrix 30–31

Earth Summit 2
ecological assessment 52–58, 178–179, 197–198; baseline studies 53; biotic indices 53; definition 52; evaluation 55–57; methodology 52–55
English Nature 53
environmental appraisal 150
environmental assessment (EA) alternatives 2; application 172–199; core values 3; costs and benefits 7; communication 12; cost 7, 9, 119–120; impact 16, 17, 25; international developments 5; methods 15–32; positive and negative lists 10, 153; principles 3, 4, 5; procedures 8, 9, 92; processes 9; techniques 33–70; terminology 3
environmental assessment methods 12, 15–32; baseline studies 18; method identification 18–28
environmental assessment techniques 18, 33–70
environmental features mapping 16, 26–28
environmental impact statements 3, 12, 116–119
environmental management systems 7
environmental statement (ES) 116–119, 127–146
environmental statement review 13, 127–146; international review procedures 134–138; procedural and technical review 128–129; review criteria 131–133; UK review procedures 129–130
environmental systems 16, 151
Espoo Convention 7
European Union 2, 3, 6, 10, 17, 62–63, 90–91, 97, 100, 125, 153, 157–158

Fifth Environmental Action Plan 157–158
financial control 119–123
Fleetwood 103–104
Flotta 141–142
France 5, 91
further reading 201
Fylde Forum 103–104

Gaussian model 37, 38–39
geographical information systems (GIS) 28, 53

Ghana 2, 3, 134
Grampian Structure Plan 162–163; council 28
groundwater assessment 61–63, 79–80, 183–184
Guri hydroelectric dam 68–71

hazard assessment 74–79; analysis 76; identification 76; procedure 75–76
Health and Safety Executive (HSE) 77, 84, 86, 88, 176
Hertz 41
Hong Kong 2, 150

Iceland 163–166
impacts 16–17, 29; monitoring 139
impacts–backwards auditing 144–146
incineration 26, 38, 78, 80–88, 123–125
indirect impacts 150
initial environmental evaluation (IEE) 10
Institute of Environmental Assessment 49, 57, 129, 133
international effectiveness study 128
ISO 14001 7

Kent County Council 161
Kent Structure Plan 161–162
Kettock Mill 28

Lake Myvatn 163–166
landfill 19, 79–80, 130
landscape assessment 45–52, 177–178, 190–192; evaluation 48–49; field survey sheet 47; impact prediction and significance 46–48
lighting 194
local planning authority 91–98
London 63

magnitude 17, 18, 21, 46, 49
Malaysia 10
Manchester review package 130–134
matrices 16, 20–25, 28–31, 65–66, 79–80, 142–144, 151; Leopold matrix 21; level (1) and (2) 21
metrolink 142–144
mineral extraction 102, 104–105
mining 101, 144–146
mitigation 25

monitoring 128, 138–146; impact monitoring 139

Namibia 2, 3, 150
National Environmental Policy Act (NEPA) 2, 5, 6, 160
National Rivers Authority (NRA) 20, 30–31, 54–55, 103, 129
National Society for Clean Air and Environmental Protection (NSCA) 62
National Vegetation Classification (NVC) 53–54, 191
Nature Conservancy Council (NCC) 7, 129, 179
Netherlands 5, 91, 128, 130, 135–136, 150, 158–159
networks 16, 25–26, 151
New Zealand 2, 5, 150
noise assessment 39–45, 181–182, 193–194; control 43–44; definition 39; methodology 40–42; terminology 41–42
North West Water 96, 103
North Western Ethylene Pipeline 57–58

oil terminals 141–142
overlays 26–28, 55

phase 1 habitat survey 53–54
Philippines 5
pig breeding 184–199
pipelines 57–58
Planning Policy Guidance Note 12 159
plume rise 37
post project analysis 13, 127–146
power plant 22–23
procedures 8, 9
professional integrity 119
programmatic environmental assessment 150, 160–161
project-level environmental assessment 151–155
project management 109–125; communication 110; context and procedure 110; financial control 119; post-mortem 123; project manager qualities 114–115; report writing 116; specialists 113; steering committee 111; study team characteristics 114–115; technical management 111

public participation 2, 12, 89–107, 135

quality assurance (QA) 119
question for thought 13, 32, 71, 88, 107, 125, 146–147, 169

Ramsar Sites 163
Red Data List 53
Redcar 141–142
references 202–215
report writing 116–119; principles 116–117
Republic of South Africa 136–138
requisite decision modelling 101
reservoir 141–142
risk assessment 73–88; applications 79–80; dermal absorption 87; estimation 76; evaluation 76–79; incidental ingestion 86–87; inhalation 82–84; terminology 75; vegetative ingestion 87–88
river corridor surveys 54–55
River Don, Aberdeen 28
roads 28–30, 152–153
Royal Commission on Environmental Pollution (RCEP) 76

scoping 2, 9, 10, 11, 16, 17, 94, 153, 176–177
Scottish Natural Heritage (SNH) 30, 53, 99, 162, 176–177, 179, 189, 197
screening 8, 9, 10, 11, 21, 25, 94, 153
sewage sludge incineration 123–125
sewage treatment works 103–104
Shell International 6
significance 17, 18, 21, 25, 48
site selection 24
site of special and scientific interest (SSSI) 10, 174, 197
Skutustadahreppur 163–166
social impact assessment 66–71; methodology 66–69
South Mayo 101
South Warwickshire 105
South West Services 96, 104
statutory consultees 91–98
steel works 141–142
stormwater runoff 63–64, 103
strategic environmental assessment (SEA) 2, 3, 149–169; basic steps 156; benefits 150–151; conducting SEA 156; cumulative

effects 151–152; key task and activities 155–156; project-level 151–154; SEA in the UK 159–160; tasks and activities 155–156; using SEA 157–161
strategic environmental assessment report 153
structure plans 161–169
Sullom Voe 141–142
surface water assessment 59–61, 183–184
sustainable development 6, 99, 150, 155
synergistic effects 151

Tarmac 96, 102
terms of reference (TOR) 110–111, 116, 122, 128
Testwood Lakes 104
Thailand 10
thresholds 151
traffic assessment 179–180, 196–197
transport planning 160
Treaty of the European Union 157

UNCED 2, 6

UNECE 7
United States 2, 5, 6, 21, 26, 59, 76, 150, 160–161
US Army Corps of Engineers 161

Victoria Falls 166–169
visual impact assessment 45–52; visual intrusion 48; visual obstruction 48

waste management 105–107, 195–196
waste to energy plant 80–88
Water Act 30
water assessment 58–64, 195–196; definition 59; evaluation 60–61, 62; standards 63
weirs 30–31
World Bank 3, 6, 12, 16, 110–111, 113, 119, 150
World Health Organisation 35, 83

Zambia 166–169
Zimbabwe 166–169
zones of visual influence (ZVI) 48, 177